RELIGION AS MAKE-BELIEVE

Religion as Make-Believe

A Theory of Belief, Imagination, and Group Identity

NEIL VAN LEEUWEN

HARVARD UNIVERSITY PRESS

Cambridge, Massachusetts

London, England

2023

First printing

Library of Congress Cataloging-in-Publication Data

Names: Van Leeuwen, Neil, 1978– author.
Title: Religion as Make-Believe : a theory of belief, imagination, and
   group identity / Neil Van Leeuwen.
Description: Cambridge, Massachusetts ; London, England ; Harvard
   University Press, 2023. | Includes bibliographical references and index.
Identifiers: LCCN 2023000862 | ISBN 9780674290334 (cloth)
Subjects: LCSH: Psychology, Religious. | Belief and doubt. | Imagination
   (Philosophy) | Group identity. | Values—Religious aspects. | Faith and
   reason.
Classification: LCC BL53 .V2825 2023 | DDC 200.1/9—dc23/eng/20230510
LC record available at https://lccn.loc.gov/2023000862

This book is dedicated to my family, friends, and teachers—
and especially to those in the intersections of those sets.

Contents

RELIGION AS MAKE-BELIEVE

Prologue

The Parable of the Playground

An extended parable can illuminate one form of religious belief, which I call *religious credence*. It can also help explain how religious credence and *factual belief* are different, though that distinction will still need clarification. My ideas about the phenomena portrayed in this parable emerge in rigorous form throughout this book, along with my arguments for them. For now, however, I want to cast a central idea that I think some part of you intuitively knows, though perhaps not in a consciously articulated way.

Imagine a group of kids who play make-believe on the playground. They get together every noon recess in the shadowy place beneath the large wooden play structure where few people can see them. There, they take out their humanoid dolls and play. The same group plays every time, and the children feel lucky to be in that group. They only let in an outsider who comes with the right kind of doll; they chase others away.

Their doll characters have names. Zalla is the strongest, capable of beating the others and calling lightning down from the sky. Hirgin is his wife; she can hear people's thoughts. Sometimes she tells those thoughts to Zalla, who reacts in mostly a just way, though sometimes his anger gets out of hand. The children imagine these doll characters each as a kind of superhero (though that word isn't *exactly* right). Each is an agent, a person-like being, who has some special power that differentiates him or her from "regular people." And no matter what happens, they can't die.

As the kids play, they sometimes have Zalla and Hirgin get into fights. Usually, they're on the same side, however, especially when there is a fight between the *other* doll characters or when people (like the kids' teacher) do something wrong that deserves punishment, like being mean. Two other doll characters are Aeter and Aul—sister and brother—both of whom wield swords and can fly. Aeter is calm and levelheaded with an ice-cold sword, but Aul has a fiery temper and a flaming sword. Ghost, their cousin, is the fastest; he flies from place to place, often riding the lightning of Zalla, and brings messages to people who need them.

The make-believe play proceeds each noon recess, as the children make voices for the dolls, build sandcastles for their palaces or forts, make the movements for the dolls as they talk or fight, and introduce other props like sticks or stones that count as swords, trees, flashes of lightning, or food. Over time, the sandcastles they've built have gotten bigger—so big that they are almost always still there in diminished form from week to week in the shadows under the large wooden play structure. Zalla's sand-castle is the biggest, and it's a point of pride among the kids that it gets repaired at least a few times each week.

The children get upset when they come outside for recess and some other group of children has already occupied their special place of play. Fortunately, after a few fights and near-fights with other kids, they've managed to gain full control over their special recess play spot, and others leave them alone. One fight was big, however. Randy and Terry from a grade above kicked down Zalla's sandcastle. Randy and Terry were older and bigger, but the kids in the group got so angry that they threw sand at Randy and Terry until they ran away. When Randy turned around and said their "stupid dolls" weren't even "real," they threw more sand at him. Two kids in the group had to go to the principal's office for this, but it was worth it; they were proud of standing up for their group and getting punished for it. After this incident, they built Zalla's castle bigger than before and continued playing.

Most days see new storylines, with different villains being punished, jailed, or killed, and with different rituals being performed. The fights between the doll characters, which are frequent, often happen because the kid who owns one doll is mad at another for some reason: if Aeter fights with Aul, it's probably because Misha is mad at Kevin for not sharing candy (or something like that); if the two doll characters make up, Misha and Kevin make up too.

The kids forget many of the events that they invent and play out, though they remember, joke about, and sometimes relive the most interesting ones. One time, when Kevin was at Cleo's house, her cat started fighting with the Aul doll and batting him around the kitchen floor. Aul got so dizzy he could barely stand up, but he ultimately got away by using his flaming sword to set the "lion's" whiskers on fire. This became "Aul and the Lion," which was frequently retold with fantastic variations. Now the kids always pretend Cleo's cat has no whiskers (even though she really has them), and whenever the kids see any other cat, they laugh because it makes Aul, who is normally tough, really scared.

Occasionally, there is a feast. One day, Hirgin used her mind powers to tell John's mom to order pizza for the kids after school. And she did! So the kids had pizza that night, and Zalla threw a feast for Hirgin the next day. Out of gratitude, the kids built up her sandcastle bigger than before—but not as big as Zalla's. (Most days, Hirgin's mind powers aren't nearly so effective, but the kids count her successes and not her failures; doing otherwise would seem disloyal.)

John, who owns the Zalla doll, is the best at remembering stories, so the other kids listen to him when he recalls what happened in the past. Sometimes, he embellishes to make things more interesting like when he said the lion bit Aul's head off and Ghost had to sew it back on (all the other kids knew it didn't happen like that but laughed anyway). But he only goes far off script when he's making a point, and he usually stays pretty close to the stories as they originally unfolded, as agreed on by the group.

John also does the setup of play for the day. He invents the dragon the doll characters have to fight or the cave they have to explore, and then the other kids play along with the situation he invents. So John is the leader, at least when it comes to their make-believe world. Because of this power, he sometimes hurts other kids' feelings by telling them Zalla or Hirgin is mad at them because of something they did. But he means well when he does it. So even though it upsets them, they usually shape up for a little while because it makes them feel better. After all, who dares suffer the wrath of Zalla?

For John's birthday, Cleo and Kevin got him a nice notebook so he could write down all the stories of what happened with their doll characters. And so he did, and he's been writing them down for some time now.

*   *   *

INTERLUDE: WHAT MAKES THIS PRETEND PLAY?

By now, you can probably see where I'm going with this. But for the parable to have its full impact, we need to investigate more thoroughly the nature of make-believe play. In particular, we need to highlight the cognitive features behind the children's activities that constitute them (in part) as *make-believe* or *pretending,* as opposed to other forms of action. Only by seeing those features clearly will we be able to notice them when they occur in less obvious ways in the setting of religious practice. The features—all of them interconnected—are *a two-map cognitive structure, nonconfusion,* and *continual reality tracking.* Importantly, all of these features are consistent with the tremendous emotive power of make-believe. I discuss each feature in turn.

Two-map cognitive structure. The children at play mentally represent *both* the <u>dolls</u> *and* the <u>mighty beings</u> for which the dolls stand.

On the one hand, the way the kids manipulate the dolls shows how they (at one level) represent the dolls as hand-sized, made of plastic, unable to self-locomote, and rigid except for a few joints that allow fixed limb movement. The various voices they make *for* the dolls reveal that they represent the plastic dolls as silent (you only speak *for* something if you represent it as not speaking on its own). This cluster of dolls-as-plastic-figure representations is part of the *first* map in the two-map cognitive structure that guides their make-believe play.

On the other hand, the kids *also* mentally represent the doll characters as mighty beings who propel themselves, have booming voices, think and feel, and are larger than typical humans. This other cluster of representations—depicting doll characters as superagents—largely constitutes the *second* map that helps guide their pretend play. Ghost can speak to Hirgin because they both—according to the second map—*have* voices (speaking for a doll character is a way of representing *its* voice). Aul can fight a lion because—second map—he's big enough to do so. And so on.

Such two-map cognitive structures characterize pretend play generally: the first map guides the pretender's movements in relation to represented physical features of the surrounding situation (among other things); the second map represents the make-believe world. Pretending, as a kind of action, requires both maps. If the first map were forgotten, the pretenders in our parable would forget to move the (plastic) dolls in requisite ways. If the second map were forgotten, the pretenders would forget the thread of the storyline.[1]

Furthermore, and importantly, *both* maps are implicated in the guidance of bodily movements in a single pretense action. Suppose Misha is playing out a dispute between Aeter and Aul. She makes a deep booming voice for Aul: "Aeter, give back my fortress!" Misha's voicing here—a single action—has a double cognitive source: because she represents the doll figure as silent plastic, she is aware that she *has to* make the voice herself if it is to be heard at all, but to make the voice for Aul, she also has to represent that *he* does have a voice; otherwise, the sounds she makes wouldn't count as *his*. Pretense actions typically have this kind of double cognitive source, where one map implies precisely what the other denies.

Nonconfusion. For the two maps behind pretend play to continue as *separate* maps, the pretending person in whose mind they exist must not confuse them (at least for most of the time as a matter of competence); if there were confusion, the two maps would collapse and become one, which doesn't usually happen. Even very young pretenders, for the most part, do *not* confuse their two maps.[2]

Many people, of course, *say* that young children take what they pretend to be real. Let's call that the Myth of Confusion. A wealth of developmental psychology shows that this myth is unfounded. One experiment is illustrative, though many could be added. Claire Golomb and Regina Kuersten had adult experimenters engage in play scenarios with young children in which Play-Doh was used to represent "cookies." The experimenter would take an *actual* bite out of the Play-Doh cookie while the child participants looked on. The children were surprised, which shows that they never took the Play-Doh for a real cookie in the first place; if they had been confused, they would have thought the experimenter was taking a normal bite of a real cookie, which isn't surprising. So, in my terms, they never confused their two maps.[3]

I believe the Myth of Confusion results partly from wishful thinking and partly from an unsuccessful attempt at articulating something true.

The wishful thinking part is that it's somehow charming to think of children as being able to inhabit a make-believe world entirely convinced of its reality, as many sentimental movies to that effect attest. But wishful thinking shouldn't influence our psychological theory construction.

What's true is that *the events of a pretend world can be emotionally significant to the children who generate them—startlingly so*. This can be seen most clearly in relation to imaginary friends. Children often feel angst or elation at what their imaginary friends do, suffer, think, or feel. But that doesn't show that the two maps are confused in their minds; it just shows that the second map—the imagined one—*also* has emotional significance.

Much evidence points in the direction of the second map's emotional significance. In her book *Imaginary Companions*, Marjorie Taylor presents the account of one of her graduate students, who as a child had imaginary friends represented by stuffed animals.

> When I traveled away from home with my family, I was allowed to take only one animal. I remember agonizing over the decision, not wanting to hurt anyone's feelings. I eventually developed a rotating system that allowed each animal to essentially go on the same number of trips as any other animal. Before each trip, I carefully selected the animal who would accompany me, and then proceeded to have a "meeting" with all of the animals together. I would tell them to the best of my ability where I was going, how long I would be gone, and what I expected to do on the trip. I reassured the animals who were staying behind that I would take them all if I could, but due to parental constraints I had been forced to choose one of them. . . . In addition to this elaborate clarification of my motives for choosing the animal that I had, I felt the need to protect the animal who was going with me from possible retaliation from the other animals upon our return. I pleaded with the other animals to be kind to the one who had been selected.[4]

The palpable emotion here is *anxiety*. The student's child-self was anxious about how her imaginary companions would react to being slighted. It's tempting to express this emotional significance by saying the companions were "real to her," but such talk is ambiguous and misleading. Saying the imaginary friends were "real to her" seems to support the Myth of Confusion, even though the phenomenon described (emotional significance) does not: emotional significance does not entail that the two distinct maps have collapsed into one another. In fact, Taylor spends most of her fifth chapter correcting the idea that children think of their imaginary companions as real, concluding that "children's mastery of fantasy is impressive. They answer many questions about imaginary objects in the same way as adults . . . and explicitly label their imaginary friends as 'just pretend.'"[5] So one should rather say, as indicated, that the second map is *emotionally significant* to the pretender (as opposed to "real to her").[6]

Nonconfusion, to return to the main thread, is evident among the children in our parable. They imagine, for example, that the doll characters' palaces and fortresses are built of marble or gold, but they always remember to repair the sandcastles that represent those palaces and fortresses *with actual sand*. They even go out of their way to find the slightly damp, *best* sand for repairing the sandcastles. This shows they never got the imagined materials (marble, gold) and the actual materials (sand) confused. And though they might *say*, "Here is the best gold!" or "Here is the best stone quarry!" their saying that is itself part of the pretending and does not indicate confusion. But the emotional significance of the sandcastles/palaces/fortresses is also evident in the way the kids angrily responded to Randy and Terry for kicking down Zalla's sandcastle, throwing sand at them until they ran away.

Continual reality tracking. We've seen that people from an early age employ a two-map cognitive structure to guide their pretend play and that they don't (typically) get the two maps confused. It's worth adding that the first map layer—the layer that, in our example, represents sand and plastic as opposed to marble and superagents—does a relatively good job of tracking basic features of reality, that it does this continuously, and that it updates to a great extent routinely in response to changes in the world. True, human minds are riddled with biases, but that shouldn't obscure the fact that successful action of any sort, including pretending, requires that one's first map layer responds to and represents, in a mostly accurate fashion, events, properties, and situations in reality that are of interest to the actor. The kids manage to meet for play because their first map layers correctly represent that the clock says noon. And they all go to the same meeting place because they correctly represent where the play place is.

They put their dolls away for safekeeping toward the end of recess because they accurately represent that recess is almost over. In short, continual reality tracking is needed for them to succeed in coordinating the *when* and *where* of collective pretending.[7]

Continual reality tracking is also an ongoing feature of pretend play itself; it doesn't just govern the start and stop. This point is implicit in my earlier description of the two-map cognitive structure, but it bears spelling out. To move the dolls successfully with their hands, the children must accurately represent the dolls' actual size, weight, and structure. When Cleo has one doll character "hide" in the "cave" she dug in the damp sand below the dry surface, she succeeds in this pretense because her first map accurately represents the doll's size and the location of the damp sand, even though her second map layer represents both the doll character and the "cave" as much bigger and more elaborate. We can generalize this point: the kids need to have mental representations that *keep them aware* of mundane features of the ordinary world to interact with those features, *even* when one is using them to create make-believe.[8]

Continual reality tracking is crucial not just to pretend play but also to representational arts generally, many of which extend make-believe.[9] Actors, no matter how immersed in their parts, keep track of trapdoors on stage, the edge of the stage, and where the audience is; without a grip on such features of reality, they would fail. And even dedicated method actors, when acting for the screen, keep track of where the cameras are. This point doesn't just extend to acting. In *The Work of the Imagination*, Paul Harris illustrates continual reality tracking in his descriptions of the ritualized burials and fantastical cave paintings that humans generated in the Upper Paleolithic.

> Cave art and ritualized burial provide clear examples . . . the artefacts and props were collectively produced and understood; they served to conjure up an imagined world distinct from the physical context in which they were manufactured or displayed. Yet, in each case that physical context *needed to be acknowledged* and re-worked if the artefacts were to serve their function.[10]

Harris's point implies a two-map cognitive structure with continual reality tracking in the first layer. Picture a person venturing into a cave to paint: that person, however preoccupied with imaginings, needs to be aware of the physical shape of the walls to do a good job of covering them with otherworldly images. And not only *do* people track reality when they're immersed in fantasy-oriented action; they furthermore *need to* do so to construct fantasy worlds effectively. The same point carries over to the physical characteristics of the playground on which the children in our parable play.

There is more to be said about pretend play. Why are the imaginings behind make-believe so emotionally salient if they're representing the unreal? What differentiates the kind of imagining behind make-believe from other kinds of imagining, like more intellectual hypothesizing or supposing? And so on. But we can already see that any action involving a *two-map cognitive structure, nonconfusion,* and *continual reality tracking* is a good candidate for being something worth calling make-believe. This is because the second map layer is not the first, and since the first can aptly be called *factual belief*—representing things like where the sand is or when recess is over—the second must be something else, something *made* over and above factual belief and not conflated with it. Still, the second map layer generates actions that often resemble the sorts of actions generated by factual beliefs—one talks to Zalla when one wants to—so in some sense, the word "belief" is apt: hence, *make-believe.*[11]

∗   ∗   ∗

THE PARABLE CONTINUED

Something unusual happened. While many playgroups dissolve when the children are still young, this one continued into young adulthood. The play with the dolls discontinued—for a time—but the kids stayed friends and frequently did normal teenage things (malls, music, etc.). Sometimes the kids would reminisce fondly about the games they used to play with Aeter and Aul and the rest, but the dolls were mostly forgotten and stayed in the attic of John's house.

One day, however, something brought the dolls back.

John's parents were getting divorced, and he was moving with his mother to a small apartment. He wouldn't get to see his father often, and he was not sure he would want to. His father had been unfaithful to his mother—or so his mom suspected but couldn't prove—but since his father had more money and better lawyers, it was clear that John and his mother would have to make do with less than they were used to. He was ashamed at school. His grades suffered because it was hard to study with his parents shouting at one another. Through all the stress, the only people he could turn to were his friends from the playgroup.

Although his friends consoled him, they caused anxiety in their own way. They all met as a group, which was frustrating because he was falling in love with Cleo. He hadn't told her, and he didn't think she noticed. Or she pretended not to notice because she was going out with Randy from the grade above, whom none of the other friends liked.

John was already agitated when he came home one day to hear his parents shouting at each other more viciously than usual. They were fighting over how to divide the furniture and other household items. John tried to intercede to get them to be more peaceful, but his father exploded.

"John, go clean out those dolls in the attic before I throw them away!"

"Dad, don't yell at me."

"Shut up! I'm tired of telling you!"

John went up to the attic in tears.

But something strange happened. A shaft of sunlight was coming through the small triangular attic window, illuminating dust particles floating through the air. John suddenly started to feel both uplifted and calm, as if the presence of the dolls took him back to a time when he wasn't scared or worried. The shouting voices downstairs receded to nothing. Everything was silent. And he suddenly felt as if a calming voice was speaking to him, saying, *You are not alone.* That feeling was so intense that he looked around to see if anyone was there, but all he saw were the boxes in the corners looking hazy and dim in the shadows.

But when he again looked toward the shaft of sunlight, he shuddered and fell to his knees. There, in the beam of sunlight in the middle of the dusty rug, was the Zalla doll standing facing him, holding a little stick in his hand, as if making ready to throw a lightning bolt to strike down the two people who were making John's life so miserable. And John felt as if he heard the voice again: *You are not alone.* For a moment, everything felt like the world was destined to be okay.

Then the moment passed.

Was the event supernatural? John didn't remember seeing the doll when he entered the attic; it was as if Zalla just appeared. And though at some level, he thought the Zalla doll must have been standing there from the last time he and his friends had played as children, the presence of the doll felt right for the moment he found himself in, like it was no accident. So, with trepidation and a new sense of purpose, he picked up the Zalla doll, put the lightning/stick in his pocket, and held the doll close to his heart. He then packed it in the box that held the other dolls and took them all downstairs where his parents had stopped fighting.[12]

John first reached out to Cleo to tell her what had happened and about the voice. She said maybe the event was a vision or supernatural visitation, and she encouraged John to tell Misha and Kevin. Also at Cleo's encouragement, John began to play with the Zalla doll again, sometimes just for fun but sometimes in an attempt to re-create that moment in the attic. When he and his mother got to the new apartment, one of the first things he did was cut out a triangle in a piece of cardboard and place it over a spot in his bedroom window so a beam of light like the one in the attic would appear on the floor. After school, he'd place Zalla in the beam of light and wait for the voice to recur; sometimes he felt a shiver, as if the voice were with him, and sometimes not. He realized that such play was unusual for a teenager, but he always felt better when he did it.

When John told Misha and Kevin, they became energized. Both felt that the time they had spent together playing in their shadowy place under the play structure had been a special time of happiness. Both wanted to re-create that time and felt that the dolls had much to do with it. Kevin even wondered aloud if the stories they had come up with as children contained hidden lessons they could use to live better lives and make the world better. Did "Aul and the Lion" actually have a deeper meaning? How could they find out?

As they discussed all this one day, Cleo made an arresting suggestion: they should return to their special place once a week and play with the doll characters to find out what the characters would do—to see if any messages would emerge. Everyone paused; it was a strange suggestion. But then each in turn agreed to try it. John tried to act calm, but he felt exhilarated.

The group chose Monday at midnight, after their parents were asleep, to meet in the sacred space on the playground. John brought all the dolls for them to play with, and he also brought the notebook that Cleo and Kevin had gotten him for his birthday many years ago, since that, too, had been in one of the boxes his father had made him clear out. The plan was just to play, and John would write down what came to him as Zalla, Hirgin, Aeter, Aul, and Ghost did their deeds and played out storylines. The first thing they did, before starting play, was rebuild Zalla's enormous palace (sandcastle) in honor of John's revelatory event that had brought them back to this place.

At first, they played at random like when they were kids. But they felt certain that the events they managed to create had deeper significance— so much so that they decided they should act out *those* events every week to show what was important to the group. John wrote down the meanings of these crucial events, though the others helped. There were many discussions (sometimes arguments) about what the events meant. John also cataloged other storylines that seemed important and interesting so they would have an official record of what the doll characters said and did. He listened to everyone for help in interpreting the events—but especially to Cleo.

After a few weeks, they settled into a pattern. They started by repairing Zalla's great palace. Then they acted out three events: the main events that had emerged over the course of their play and the ones during which John had sensed the voice's return.

First, Zalla and Hirgin hold hands at the top of Zalla's palace. This means *Love.*

Second, Ghost takes a lightning bolt as a message of justice from Zalla and flies down from the top of Zalla's palace and into the world/playground. This means *Truth.*

Third, Aul and Aeter lay down their swords. This means *Peace.*

After the ritual playing out of these three events, play would proceed boisterously as usual. And over time, the group began to call themselves The Playground.

The Playground group slowly grew in number. As the founders told trusted friends, the number of regular attendees grew to about a dozen in the course of a few months. The success of the group was largely due to its Three Principles, commitment to which defined the group, even as the principles "lifted them up," as they commonly put it, in their lives. The Three Principles were

1. *You are not alone.*
2. *Love, Truth, and Peace are gifts from above.*
3. *No matter how bad things get, you can always come back to this place.*

These principles were attractive to the teenagers as they went through tumultuous transitions in their lives. The first was Founder John's message from Zalla. The second comes from the actions that had been revealed to the group when it was just (re-)starting. The third is the assurance that The Playground would always be there for those committed to it.

And the principles, supernatural beings, and rituals stayed attractive to the members as they grew up.

Founder John and Founder Cleo eventually got married and started small Playground Communities in several cities. Founder John used donations from members of The Playground to support himself while he wrote the *Book of Powers,* which contained the stories, lessons, and principles that had come to them as a group. *Powers* was the term the four founders eventually decided on for the supernatural beings Zalla, Hirgin, Aeter, Aul, and Ghost. Founder Kevin and Founder Misha read John's drafts thoroughly and made extensive comments, but his most important sounding board was Founder Cleo, who was both critical and supportive. Collectively—through suggestion, discussion, extended play, and argument—they worked out how to present the meanings of John's initial vision in the attic and his subsequent experiences in a way that would be most impactful to members and potential members. They decided to write down, for example, that Founder John had been *entirely certain* that the Zalla doll had *not* been standing there when he first went into the attic (a rendering to which John reluctantly agreed[13]). Furthermore, though Founder John had initially said it was unclear to him whether the voice he "heard" had been Zalla's own voice or merely a voice associated with Zalla, the group eventually decided to equate the voice with Zalla for purposes of the book.[14] Thus, the *Book of Powers* was a collective enterprise, even though Founder John was counted as the sole author and visionary.

As The Playground grew, communities appeared all over the nation in YMCA gyms, in living rooms, in rented-out church rooms, in black box theaters on off days, and on school playgrounds. Most groups were peaceful and supportive, but those that insisted on doing *all* their playing on actual school playgrounds (no other settings allowed!) became known as

The Fundamental Playground. Sadly, that branch began doing alarming things. They would use their consecrated playgrounds even after school districts had ordered them to leave, and, on occasion, groups of them would even fight with the police officers who had been sent to remove them—viciously throwing sand at the officers. Those who enacted or suffered violence for the Fundamental Cause, as they called it, were hailed as "lifted up." And enormous sandcastles would continue to appear on consecrated playgrounds, long after the Fundamentals had been forced to leave. Most disturbing, some Fundamental members were caught destroying the lightning rods on "enemy" buildings in an effort to call forth the wrath of Zalla, an act they thought was licensed by the "true" and "strict" interpretation of the *Book of Powers.* Nevertheless, as indicated, most Playground communities were not nearly so extreme, though individual members differed widely in their level of sympathy with the Fundamentals.

John is old now. But he still has his original Zalla doll, replicas of which appear in every building in which mainstream Playground communities "play" in many cities. And no matter where he lives—he and Cleo continue to move from city to city—he finds a room in each new house or apartment through which sunbeams shine. He still makes a cutout or other manner of triangle shape from cardboard or wood each time so that he can place Zalla with his lightning bolt in the beam of sunlight once a day. If he is lucky, he can feel the presence of a voice saying to him, *You are not alone.*

WHAT THIS BOOK IS ABOUT

This book is about the broad class of psychological states called "beliefs," as well as the relation between "belief" (of various sorts) and *imagination.* In time, we will transition from focusing on such cognitive elements to more emotion-laden and value-driven components of religious psychology: in particular, group identity and sacred values.

I started with this parable because it captures a conceptual possibility for how to think about religious belief that has been largely overlooked in contemporary philosophy of mind and cognitive science. Roughly, the possibility is this: many religious beliefs are *imaginings* of the sort that guide make-believe play, though they are imaginings that become central to the religious actor's identity and guide symbolic actions that express sacred values. My view is that this conceptual possibility *in fact* describes many so-called religious beliefs in the minds of actual people around the world.[15] If that's true, the consequences are far-reaching.

The class of "beliefs," from my point of view, is *not* as unified as many philosophers glibly suppose (hence my frequent use of scare quotes). Still, there are deep and interesting patterns in the variation, and one way to describe the aim of this book is as an effort to articulate those patterns and situate them within a broader, systematic theory of cognitive attitudes—mental states that represent how the world or some portion of it *is* or *might be*. Is a typical Playground member's "belief," for example, that Zalla throws lightning the *same* sort of mental state as her "belief" that lightning is an electrical discharge? Is it slated to receive the same sort of processing in her mind? I doubt it, and the differences that exist cannot entirely be accounted for in terms of differences in content; rather, the two "beliefs" coexist in different map layers in the member's mind that interact in interesting ways. And the distinct map layers have different *attitudes* attached to them. If I am right about this, and if the psychological processes illustrated in the parable are plausible, then we owe it to epistemology, psychology, cognitive neuroscience, anthropology, religious studies, and history—not to mention public discourse—to find the fault lines in the category of "belief" and articulate them.

To put these points more formally, my two main theses, which will receive much clarification as we progress, are these:

> **Distinct Attitudes Thesis**: *factual belief* and *religious credence* both exist and are distinct cognitive attitudes (they are two different ways of processing ideas).

> **Imagination Thesis**: religious credence differs from factual belief in many of the same fundamental ways that *fictional imagining* does—by "fictional imagining," I mean the cognitive attitude that underlies pretend play.[16]

Just as there was a two-map cognitive structure, nonconfusion, and continual reality tracking in the psychological processes underlying the *pretend* actions in the first part of the parable, there was also a two-map cognitive structure, nonconfusion, and continual reality tracking in the psychological processes underlying the *religious* actions in the second part: as time went on, fictional imaginings morphed into religious credences while remaining distinct from factual belief. Just ask: Were the plastic figures *dolls for play*, or were they *idols for worship?* They were clearly both. The parable thus illustrates the continuity between fictional imagining and religious credence as vividly as possible. But if I am right—and this claim will strike many as radical—two-map cognitive structures are not only characteristic of the practicing religious minds in the parable; they are also characteristic of many (and maybe most) religious minds in actual people around the world. Factual beliefs about whales and religious credences about how a prophet named Jonah spent three days inside one receive systematically different manners of psychological processing—differences, again, that can't be accounted for *merely* by appeal to differences in contents. Thus, another major departure from most research on religious "beliefs" is a shift in focus from the *contents* of those beliefs to the *attitude* that is taken toward them: understanding this overlooked dimension clearly will illuminate how, why, and to what extent much religious practice is pretend play.

None of this is to say that there are no differences between *mere* imagining and religious credence. As I argue in Chapter 6, religious credences play distinctive roles in constituting people's group identities; relatedly, religious credences connect people's sacred values systems *to* entities in the world that thereby come to be regarded as inviolable in a sense I explicate in Chapter 7. But all that is consistent with the fact that religious credence is far different from factual belief—and much more like imagining—in its cognitive dynamics. Religious credence, very roughly stated, is imagining *plus* group identity and sacred value.[17]

My main work is theoretical. Nevertheless, the theory I develop is constrained by current empirical evidence (such as it is[18]) and is meant to foster further empirical research in turn—which will then motivate revisions to the theory, and so on. Furthermore, our view of the class of phenomena that constitutes our subject matter will shift as the theory develops, as is usual for research programs that describe natural phenomena.[19] Religion, in any case, is a multifarious category, as Pascal Boyer is fond of pointing out.[20] So how widely my theory will extend is an open empirical question; it would be folly to judge particular cases before the detailed evidence is in. For example, does my notion of *religious credence* apply in the same way to conservative Catholics in Spain as it does to devotees of spirit animals in Amazonia? We don't know without first understanding the theory and applying it carefully to the gathered evidence. But we can't even pose such questions properly without the kind of disambiguation of the notion of "belief" that my theory offers. In other words, my goal is not to argue that every mental state in the world that can be pretheoretically called a religious "belief" fits my characterization of religious credence—rather, my goal is to provide the expressive power to identify and explain important differences where differences exist. We'll soon see that this goal alone brings startling consequences.

CHAPTER ONE

The Attitude Dimension

1. WHERE SHOULD WE START?

The Playground religion was an illustration of the difference between factual belief and religious credence. While having a factual belief that a plastic doll is atop a sandcastle, for example, John has the religious credence that mighty Zalla is in his enormous palace. There are two cognitive maps in John's mind.

But what, exactly, *are* factual belief and religious credence? In philosophical parlance, they are *attitudes*. To understand what this means, we need to start with the attitude/content distinction, which is widely used in philosophy of mind and other parts of cognitive science. So this chapter is somewhat technical, but it will enable us to focus on the important yet undertheorized dimension along which many religious "beliefs" differ from mundane factual beliefs: *the attitude dimension*. Once we understand this dimension, subsequent chapters will fill in the details of what constitutes the differences between these attitudes and many related ones.

This chapter thus aims to foster an ability: the ability to think about the attitude dimension of psychological states independently of other dimensions. This ability is essential to understanding everything that follows. It is also not trivial: in Section 4, we'll see that even trained philosophers of mind fall into confusion by not applying the distinction properly. Yet seeing their mistakes will illuminate a way forward: attending carefully to the attitude/content distinction opens up surprising and interesting consequences.

2. THE ATTITUDE/CONTENT DISTINCTION

Consider four reports one might make about people's mental states:

> 1) Kevin <u>doubts</u> *that there is a possum in the bush.*
> 2) Misha <u>hopes</u> *that there is a possum in the bush.*
> 3) Randy <u>doubts</u> *that God exists.*
> 4) Terry <u>hopes</u> *that God exists.*

The underlined words refer to the people's <u>attitudes</u>. The italicized words indicate the *contents* of their mental states. The attitudes are the different ways the people process their respective ideas—in this case, a doubting attitude versus a hoping attitude. The contents (roughly) are what those ideas are about.[1]

Those two dimensions of mental states move independently of one another. Kevin and Misha represent the same content (*that there is a possum in the bush*), but they have different attitudes toward it (doubting versus hoping): they relate to that same content in different ways. Similarly, Randy and Terry both mentally represent the same content (*that God exists*) while relating to it in the same two different ways (doubting versus hoping).[2]

So an attitude is *a way of processing* ideas: for any given idea, one can process it in a <u>doubting</u> way, a <u>hoping</u> way, a <u>suspecting</u> way, an <u>assuming for the sake of argument</u> way, an <u>accepting for practical purposes</u> way, and so on.

This is a somewhat different use of the word "attitude" from what you're likely to find in social psychology. In social psychology, "attitude" often refers to a broad outlook on some social group or issue—an outlook that nebulously includes various components: evaluative, descriptive, emotional, biases, unconscious behavioral tendencies, and so on. But the way I'm using "attitude" (which is common in philosophy and cognitive science) is more targeted in that it encompasses particular ways that people process specific ideas rather than being a broad orientation concerning a suite of topics. For example, if you say Terry <u>assumes for the sake of argument</u> that Goldbach's conjecture can be proven, you've described an attitude—the attitude of assuming an idea for the sake of argument.[3] That said, attitudes in this specific sense are in another way more general: it is not just social situations, groups, or issues that can be the target of attitudes; rather, one has attitudes *in relation to anything whatsoever that one thinks about* (from what one's cat is named to Goldbach's conjecture). For any idea that you have, you're processing it *somehow,* which—on the usage I stick to henceforth—means you have one attitude toward its contents or another.[4]

So for any given attitude, there is no proprietary content that can or can't go with it. For any content *p* (religious or naturalistic; scientific or commonsense; concerning the observable or the unobservable), you can in principle have any attitude toward it. True, some ideas tend to go with some attitudes more than others; for example, people are much more likely to <u>fictionally imagine</u> ideas about unicorns than they are to factually believe them. But someone who was misinformed about zoology could also factually believe that unicorns exist.[5] So a mental state's content, though heuristic, is never a decisive indicator of its attitude type (we will see how thinking that it is leads to mistakes). The independence of attitude and content is thus an important fact about human cognitive flexibility. To help internalize this flexibility, practice visualizing with my <u>underlining/</u> *italics* notation: imagine people's different <u>attitudes</u> mixing and matching with various *contents* and see if any combinations jump out as unexpected (e.g., Jeremy <u>factually believes</u> *that unicorns exist*).

Philosophers generally divide attitudes into *cognitive* and *conative.* Cognitive attitudes, roughly, are those that present the world as *being* a certain way: when a person has a cognitive attitude that *p,* she is disposed in some way and to some extent to think and/or act as though *p* is true.[6] Cognitive attitudes include <u>factually believing</u>, <u>supposing</u>, <u>fictionally imagining</u>, <u>hypothesizing</u>, <u>assuming out of caution</u>, and so on. Conative attitudes, roughly, are those present some content as an *aim*; when a person has a conative attitude that *p,* she will be inclined to do what it takes to bring it about that *p* if she has the chance. Conative attitudes include <u>desiring</u>, <u>wishing</u>, <u>hoping</u>, <u>longing for</u>, and so on. As we'll see in more detail in the next chapter, it is the combined psychological work of cognitive and conative attitudes that makes choice and action possible.[7]

With that machinery in place, we can now sharpen our focus. The questions are these: What is the cognitive attitude that members of The Playground have in relation to their first map layer? I've labeled it factual belief, but much remains to be specified. Still, we know what we need to figure out: How is it that they are *processing* that first map layer? And how shall we characterize the attitude they take in relation to their second map layer? I've called that religious credence, but that, too, leaves much as yet unspecified: How exactly is the religious credence manner of processing different from the factual belief manner of processing?

Even this much, however, lets us highlight an important issue with the words "believe" and "belief." In the same way that the word "star" can nonscientifically apply to two distinct astronomical phenomena (visible planets like Venus and burning balls of gas like Alpha Centauri both get called "star"), so, too, can the pretheoretic word "believe" apply to distinct psychological phenomena: to (at least) two very different cognitive attitudes.[8] This is why researchers often introduce new terms (*terms of art*) for the sake of talking about distinct phenomena more clearly and avoiding the conflations of pretheoretic speech. Consider the following attitude reports spoken casually:

> 5) Jane believes[#] *that John Madden is alive.*
> 6) Fred believes[*] *that Jesus Christ is alive.*

Although the attitude verb in these reports is the same ("believes"), that might be masking two different ways of relating to the idea that someone is alive. Jane probably missed the sad news on December 28, 2021, and thus thinks, in a matter-of-fact way, that John Madden still lives (and note that "thinks" rather than "believes" would be more natural for the Madden attitude report, as we'll see in Chapter 5). But Fred is more likely to be taking a reverential, identity-constituting attitude toward the idea that Jesus lives. So 5) and 6) could be more exactly and technically rendered:

> 5′) Jane factually believes *that John Madden is alive.*
> 6′) Fred religiously creeds *that Jesus Christ is alive.*

Of course, I still need to do much to give that distinction detail and to argue that it has real-life application (that work will come in Chapters

2–4). But now that we understand the attitude/content distinction, we can see what dimension of psychological states is at issue.

*   *   *

Before moving on, let me give a further technical note that will be useful later, though it can mostly be bracketed without a loss of understanding of my main points. The attitude/content distinction falls within a broader framework, one that distinguishes *four* components of mental states: agent (who has the mental state), relation or attitude (what the manner of processing is), mental representation (the specific structure by which the mental state represents something), and content (what the truth or accuracy conditions are of the mental state). Within this framework, one can describe many mental states with an ordered quadruple: **aRmc** (**a**gent, **R**elation/attitude, **m**ental representation, **c**ontent). We won't need to think in terms of every component of the broader framework for most purposes throughout the book because the attitude dimension is what's under consideration, and that can be largely grasped with the attitude/content distinction alone. Still, there will be some places in the book where representational structure and the distinction between *that* and content will be worth keeping in mind—especially in Chapter 8—which is why I introduce the broader framework here.[9]

3. APPLYING THE DISTINCTION

We now return to my two main theses, introduced in the Prologue, since we can understand what they imply and don't imply with much greater clarity in light of the present discussion. Once again:

Distinct Attitudes Thesis: *factual belief* and *religious credence* both exist and are distinct cognitive attitudes (they are two different ways of processing ideas).

Imagination Thesis: religious credence differs from factual belief in many of the same fundamental ways that *fictional imagining* does—by "fictional imagining," I mean the cognitive attitude that underlies pretend play.

In short, much as fictional imagining is a different way of processing an idea from factually believing it, so, too, is religious credence a different way of processing an idea from factually believing it.

If I've explained this framework well enough, the following methodological points may be obvious. But let's spell them out explicitly to avoid common pitfalls.

First, since *religious credence* and *factual belief* are terms of art for different <u>attitudes</u>, neither term implies that any given religious credence or factual belief is true or false. What makes a cognitive attitude true (or false) is the relation between its *content* and reality (is reality as the content describes?). For example, one might have religious credence with the following content: *a baby named Jesus was born in a town called Bethlehem about 2,000 years ago.* It is possible, at least in principle, to have a factual belief with that same content. The truth or falsity of *both* of those mental states will stand or fall with whether that content accurately describes what happened; the attitude type does not determine truth or falsity one way or another. More generally: a religious credence with p as its content and a factual belief with p as its content both depend on that content (p) for truth or falsity—the attitude type doesn't guarantee truth or falsity either way.[10]

Second and relatedly, since attitude type and content come apart, it is possible for some religious credences to have (in a different use of the word) "factual" contents. It is also possible for some factual beliefs to have "religious" contents. As an example, consider the content *that Stalin's policies did not cause famine in Ukraine.* That content concerns what might be called matters of fact since it is historically true or false (here, it's false since Stalin's policies did cause famine—and did so deliberately). But is it a factual belief in the sense I'm developing? Trick question. Since it's a content, one can in principle have the attitude of factual belief *or* the attitude of religious credence toward it (and many other attitudes besides). A history student in Cleveland who misread her textbook might arrive at the following mental state:

Janna <u>factually believes</u> *that Stalin's policies did not cause famine in Ukraine.*

Janna has an attitude of factual belief held in relation to false contents. She is, so to speak, merely mistaken. But an ideological fan of Stalin who has reached a certain level of fervor might well enter the following mental state:

> Kai <u>religiously creeds</u> *that Stalin's policies did not cause famine in Ukraine.*

Despite having the same contents, the psychological dynamics of Janna's and Kai's mental states are very different. Janna's will much more easily update in light of evidence (e.g., if she more carefully reads her textbook), among other significant differences.[11] Importantly, the fact that the attitude of religious credence can be taken on topics that concern matters of fact helps explain why religious dogmatism can infect debates that, from a content perspective, should be settled scientifically. In any case, keep in mind that the phrases "factual belief" and "religious credence" are terms of art for cognitive attitudes. One can *also* use the words "religious" and "factual" for many other purposes, but those other uses shouldn't be confused with the specific attitude notions I'm defining.

Third, since any given attitude type is composed of multiple functional features (whether it is under voluntary control, whether it updates in light of evidence, how it influences reasoning, etc.), we should not see differences in attitude type as sharp and rigid divisions. Rather, we should think of attitudes as being clusters within a property space, where certain features systematically tend to go together, but not as a matter of necessity. Otherwise put, attitude types form *attractor positions* within psychological space.[12] There will be straggler mental states that fall outside the clusters, but those are interesting exceptions, and the theory that follows will allow us to better describe them too.

4. A CASE OF ATTITUDE/CONTENT CONFUSION

Most of what has been said in this chapter is familiar to philosophers of mind and many cognitive scientists. Yet it is surprisingly easy for those trained in these fields to make mistakes by not attending to the distinctions just explained. Consider this endnote from Nicolas Porot and Eric Mandelbaum's recent review article on the cognitive science of belief, which exemplifies a mistake I've seen elsewhere as well[13]:

A problem for Van Leeuwen style accounts is that factual beliefs also seem to have problems updating. . . . The question of whether man-made global warming exists is a factual question, and one that leads to a factual belief, yet the updating of it is still stubbornly recalcitrant in the face of evidence.

Here, they are responding to my position, which I have discussed before and elucidate in Chapters 2 through 4, that one of the things that differentiates religious credence and factual belief is that factual beliefs involuntarily update in response to evidence, while religious credences do not.[14] Porot and Mandelbaum are saying that that position isn't right because many "factual beliefs" (like those about global warming among some people) also don't update in response to evidence.

The discussion of Janna's and Kai's mental states from the last section should reveal the mistake. In objecting to my position that factual beliefs update in light of evidence, Porot and Mandelbaum have picked out a *topic* (or kind of *content*) that one might call "factual" and assumed that having in mind that sort of *content* (about global warming or whatever) makes any "belief" that has it a "factual belief" as an <u>attitude</u> type.

Having grasped the attitude/content distinction, you can see why that move is a nonsequitur (and, indeed, it is a nonsequitur on *any* theory that draws an attitude/content distinction, not just my own). Since any content can go with any attitude, one can have all sorts of different cognitive attitudes about some "factual" content, like *that anthropogenic global warming exists.* Consider the following attitude reports:

Jeff <u>hypothesizes</u> *that anthropogenic global warming exists.*
Sarah <u>wonders whether</u> *anthropogenic global warming exists.*
Dillon <u>assumes for the sake of argument</u> *that anthropogenic global warming exists.*
Sam <u>doubts</u> *that anthropogenic global warming exists.*

None of these reported mental states—which are common enough—are factual beliefs in the attitude sense in question, even though their content concerns a "factual question." So Porot and Mandelbaum are wrong to infer factual belief as an <u>attitude type</u> from a mental state's "factual" content: if that move were legitimate, all the mental states just mentioned would be factual beliefs, but none of them are (they are, rather, <u>hypothe-sizing</u>, <u>wondering</u>, etc.).

Despite its nonsequitur, Porot and Mandelbaum's passage does point to something important at which I've already gestured: in the case of anthropogenic global warming, many people's attitude type is (sadly!) *not* factual belief in the relevant sense. If it were, their "beliefs" would update far better than they do. More specifically, many denialists have an attitude similar to religious credence with the content *that there is no anthropogenic global warming.* Call it <u>ideological credence</u>.[15] Thus, the mental states that Porot and Mandelbaum gesture at have the following structure:

S <u>ideologically creeds</u> *that there is no anthropogenic global warming.*

Such "beliefs," rather than being a challenge to my view, are *better* described by an extension of it than by frameworks, such as Mandelbaum's, that would erase the distinctions I draw.[16] Note, by way of contrast, that many people's beliefs about global warming *do* update over time with evidence.[17] In those cases, people's cognitive attitudes about global warming most likely are or were factual beliefs. So, importantly, my framework has the expressive power to describe crucially different mental states, even when they concern the same contents. One can have ideological credences *or* factual beliefs about global warming; those are psychological states with different dynamics. Respecting the attitude/content distinction enables us to characterize those differences with clarity and avoid the mistake that Porot and Mandelbaum (and others) make.[18]

5. ANYTHING CAN BE SACRALIZED

This chapter clarifies how to think about the difference between the map layers in the two-map cognitive structure in the minds of The Playground members. The attitude/content distinction clarified the dimension of mental states along which religious credence and factual belief differ: the attitude dimension—how the different map layers are processed.

In getting clear on this dimension and its independence from other dimensions, we came to an important realization as a theoretical bonus, one that should resonate with recent experience in public life. That realization can be put as follows:

Anything can be sacralized.

That is, the attitude that humans employ in rendering a certain entity, topic, idea, or proposition sacred needn't be exclusively attached to super-natural subject matters (though there are systematic reasons why it often is, as we'll see in Chapters 6 and 7). Rather, through the attitude of religious credence, we humans possess the capacity to make symbols out of any subject matter—for better or worse. To extend that point to the realm of action—just as one can play games of make-believe concerning any topic—so, too, can one play sacralized games of make-believe concerning any topic. What this chapter has done, I hope, is enable you to think clearly about that possibility.

CHAPTER TWO

A Theory of Cognitive Attitudes

1. A NEEDED THEORY

David Hume once asked an underappreciated question: "Wherein . . . consists the difference between . . . fiction and belief?"[1] Even though he used terms different from my own, he was asking about the difference in *attitude type*.

Hume's point was that it's difficult to explain what distinguishes merely imagining an idea from actually believing it (factually believing it, in my terms). To use the notation from the last chapter, the question is: What are the constitutive differences between these two mental states?

1) Sam <u>factually believes</u> *that p*.
2) Sam <u>fictionally imagines</u> *that p*.

Many superficially plausible answers to this question end up being circular—or just wrong.[2] One might say, for example, that 1) involves "regarding as real," while 2) involves "regarding as fictional." But what is "regarding"? It looks like another cognitive attitude term (somewhat disguised) that refers to having a *belief* of some sort, which is the very thing we were supposed to explain. The supposed answer here just duplicates the problem. Hume's answer was that belief is a sentiment that is more "vivid, lively, forcible, firm, [and] steady" than mere imagining. The problem with this is that many beliefs don't *have* a vivid or forcible sentiment, like my matter-of-fact belief that there are pens in my drawer. And a person's imaginings can be vivid and forcible, like an arachnophobe's imagining of a spider under his bed. So Hume's question was better than his answer to it.

What, then, does differentiate factual belief from fictional imagining? Answering this is central to explaining how humans can be in touch with reality—and get it right or wrong. And it is especially important for this book because the insights we glean in answering the question will help differentiate matter-of-fact cognition, involving factual belief, from religious cognition, involving religious credence (which shares fundamental features with fictional imagining).

To that end, the theory developed in this chapter answers Hume's question and also characterizes other cognitive attitudes in addition to factual belief and fictional imagining, such as hypothesis, supposition, assumption for the sake of argument, acceptance in a context, and so on, which, along with fictional imagining, I call *secondary cognitive attitudes* for reasons that will become clear. Section 3 gives the theory in full, but before we get to it, I want to set up goalposts to help us evaluate the theory's success.

2. HUME'S, QUINE'S, AND CLIFFORD'S DESIDERATA

A desideratum is a pair of goalposts: a theory that satisfies a desideratum has accomplished an important goal. This section posits three desiderata for theories of "belief" and other cognitive attitudes. At the end of the chapter, I say how my theory satisfies them. But they aren't useful only for judging my views; you should apply them to *any* theory that purports to be about belief.

The first desideratum is that a theory of belief or imagining (or any cognitive attitude) should say what distinguishes one cognitive attitude from the other. It should, for example, distinguish factually believing an idea from merely imagining it fictionally (or imagining it for some other purpose, like practice). I call this Hume's Desideratum in honor of Hume's question.

Hume's Desideratum: a theory of belief (factual or otherwise) should explain what distinguishes belief from other cognitive attitudes, like fictional imagining.

Let's say Kevin from the Prologue mistakenly thinks there are apples in the fridge. He doesn't merely (fictionally) <u>imagine</u> there are apples in the fridge; he <u>factually believes</u> it (which is why it counts as a mistake in a way that fictionally imagining false contents does not).[3] But what constitutes the difference? Or, what's the difference between <u>hypothesizing</u> there are apples in the fridge and <u>factually believing</u> it? Hume's Desideratum states that a theory of belief is lacking if it doesn't supply answers to these questions.

Twentieth-century philosopher W. V. O. Quine inspired the second desideratum. Quine thought that metaphysics should be continuous with the empirical sciences like biology, chemistry, and physics. For our purposes, that means that philosophy of mind should mesh with empirical psychology. Our ultimate theory of factual belief, imagining, and so on should connect research in philosophy *to* empirical sciences like psychology and neuroscience. Quine, admittedly, was a behaviorist who sympathized with his friend B. F. Skinner, so he would have objected to the internal mental representations and processes my theory posits. But such posits are now standard in cognitive science, so we must choose between Quine's behaviorism (which rejects mental representations) and his principle of continuity between metaphysics and science (which would bring them back, given contemporary cognitive science). I choose the latter option. So we have:

Quine's Desideratum: a theory of (factual) belief and other cognitive attitudes should unify philosophical and psychological research on those topics.

On this front, both philosophers and psychologists have been remiss. Philosophers who talk about "belief" often ignore empirical data—consider most analytic epistemology and philosophy of religion—while many psychologists use the term "belief" without clarification, which often makes their claims hard to interpret.[4] Quine's Desideratum pushes us to apply the systematic conceptual clarifications of philosophy *to* the evidence-based claims of psychology and vice versa.

Our third desideratum is named after William Kingdon Clifford, the nineteenth-century English mathematician and philosopher. Clifford argued that it was immoral to hold beliefs without sufficient evidence. His key example was someone who believed without evidence that a ship was safe to sail. To hold such an evidence-lacking belief would be immoral, says Clifford, because it would lead to behaviors that endanger people.[5] My aim here is not to assess Clifford's claim either way. Rather, Clifford's claim is an example of an important phenomenon: *people, as a matter of fact, often hold beliefs of various sorts to normative standards.* People criticize or evaluate other people's beliefs. We argue over who is right, where being "right" often means having true factual beliefs. We give *reasons* for what people should think. A theory of beliefs should help explain why people get so worked up, normatively speaking, about them—much more so than about other attitudes, like imagining. Clifford wouldn't say there's something wrong about merely *imagining* without evidence that the ship is safe to sail. Hence:

> **Clifford's Desideratum**: a theory of (factual) belief should help explain why people hold beliefs to more stringent norms of rationality than they do for other cognitive attitudes.

3. THE THEORY ITSELF

Consider your drive home from work. To get home, you rely on factual beliefs. You factually believe your address is 123 Smith Lane. You have factual beliefs about what streets lead to Smith Lane. You factually believe a certain key on your chain is the car key. You have factual beliefs about what each dashboard gauge indicates (that's the gas level, that's the RPM, etc.). You factually believe that the red octagon sign requires you to stop. Most of the factual beliefs you have count as knowledge, though not all.

For the remainder of this chapter, all the *beliefs* that I talk about are of this matter-of-fact variety. For this stage of theory construction, I'm setting aside beliefs that come in degrees (see the endnote); moral beliefs; and, of course, religious "beliefs"; to which we'll turn in the next chapter.[6] Once we've adequately theorized obvious cases of factual beliefs, we can apply this theory to those other psychological states—those "beliefs"— in illuminating ways to see where they fall within the *space* of cognitive attitudes.

To start, whenever you say that someone *thinks* or *believes* that *p* (in the factual belief sense), there is an implied contrast. You're saying that person does something *more* (cognitively speaking) than merely imagining that *p*. But what is the "more"?

The most accepted theory of belief in philosophy today doesn't explain the "more." On that theory, advocated by both Hume and Donald Davidson, beliefs are mental states that cause actions together with desires.[7] What causes you to turn left on Smith Lane? First, you <u>want</u> to get home. Second, you <u>believe</u> your home is on Smith Lane. The Hume–Davidson theory, which forms the basis for contemporary decision theory, says beliefs are mental states that cause actions (turning left) that would help satisfy desires (your desire to be home) in situations in which the relevant beliefs are true (your house really *is* on Smith Lane). Belief is the mental state that plays *that* role in generating action.

But David Velleman points out that other attitudes, like imagining, sometimes play that role too. Kevin, for example, <u>imagines</u> his teddy bear is Santa, <u>wants</u> to tell Santa his Christmas list, and thus *speaks out* his Christmas list to his teddy.[8] Here, a <u>desire</u> + <u>imagining</u> pair causes the same outward behavior (at least on a coarse view of the behavior) that a <u>desire</u> + <u>belief</u> pair would cause (that is, the <u>desire</u> + <u>imagining</u> pair causes his speaking out the Christmas list). So, ironically, the Hume–Davidson theory of belief doesn't satisfy Hume's Desideratum of distinguishing belief from imagining.[9] Their theory is not wrong, exactly, but we do need a finer specification of belief than the theory gives if we are to distinguish belief from imagining.[10]

The theory I give below is designed to specify the "more"—to say how factually believing a proposition is *more* than merely imagining it. Here are the principles that form the backbone of my theory, along with terms for the properties they describe:

1. *If you factually believe it, you can't help believing it.*
   (I call this **involuntariness**.)
2. *Factual beliefs guide action across the board.*
   (Let's call this **no compartmentalization**.)
3. *Factual beliefs guide inferences in imagination.*
   (This is **cognitive governance**.)
4. *Factual beliefs respond to evidence.*
   (Let's call this **evidential vulnerability**.)

Let's motivate these with some intuitive examples before clarifying and defending them with greater precision in the following subsections.

For involuntariness, consider this: you can't choose, through an act of *will,* to factually believe the current year is 2004 or that you're wearing a birdhouse for a hat (just try—what would that even amount to?). Nor could Kevin from the Prologue simply choose to believe that the sand around him was gold, though he could easily choose to *imagine* that. That means his second map layer had a voluntary latitude that his first one didn't. In the words of Dan Sperber, from the point of view of the believing subject, "factual beliefs are just plain 'knowledge.'"[11] And you can't voluntarily make something seem like just plain knowledge if it doesn't, nor can you voluntarily just decide to stop believing something that seems to you like just plain knowledge. One can, of course, voluntarily do things that *indirectly* change one's factual beliefs, like looking up information in a book. But you can't choose to change your factual beliefs directly in the way you can choose to raise your arm—or in the way you can choose to imagine something. If you factually believe it, you can't help believing it.[12]

No compartmentalization means that factual beliefs guide people's actions in all situations. Consider Kevin again. On the playground, he factually believes he's surrounded by sand, that the sand underneath is wet, and so on. These factual beliefs continue to guide his behavior, even when he's playing pretend or when he's in different settings altogether.[13] In science class, he factually believes one object is a magnet and the other a fork, and these factual beliefs guide how he tests his hypothesis about magnetizing a fork. In gym class, he factually believes he can't touch the rim of the hoop, which is why he imagines dunking to motivate practice. In every situation in which he acts, at least *some* of his factual beliefs (the relevant ones) supply cognitive inputs into action choice: in other words, factual beliefs are not compartmentalized to one practical setting or another. In contrast, secondary cognitive attitudes *are.* On the playground, Kevin <u>imagines</u> Hirgin getting angry with Zalla—then he acts this out by making pretend voices. But such imaginings are compartmentalized to the make-believe setting; they stop guiding behavior outside it.[14] In science class, he has <u>hypotheses</u> that he sets aside when class is over; hypotheses are thus compartmentalized too (to the setting of investigation). In gym class, he <u>imagines</u> dunking a basketball to practice jumping, but that imagining stops guiding his behavior once he's out of that setting of practice. So imagining and hypothesis guide Kevin's actions in specific settings, but those attitudes are compartmentalized. In other words, secondary cognitive attitudes have "off" switches: they stop guiding behavior outside their special practical settings. In contrast with compartmentalization, factual beliefs as a class guide action in all settings—across the board. That's because factual beliefs, from your standpoint, do the work of portraying how things really are—things that *must* be dealt with, if you confront them or they confront you, in any setting.

Cognitive governance means that factual beliefs guide how imagining unfolds. When Kevin <u>imagines</u> lightning hitting a tree, he next <u>imagines</u> the tree bursting into flame. But that imaginative transition—from imagining the lightning strike to imagining the flames—doesn't come from nowhere. Kevin's <u>factual beliefs</u> (that lightning is extremely hot and that trees are wood and hence flammable) guide his imagination from one imagining to the next. If he didn't have these factual beliefs about lightning and trees, his imaginings wouldn't unfold in that fashion. Thus, factual beliefs *govern* inferential transitions among imaginings (as well as other secondary cognitive attitudes). Otherwise put, the information stored in a person's factual beliefs guides how that person's imaginings unfold; it does the same for other cognitive attitudes. Importantly, imaginings don't do the same for factual beliefs, and this lack of symmetry is a defining contrast.[15]

Evidential vulnerability both characterizes factual beliefs and helps explain our success in the world. When Kevin is hungry, he goes to the fridge, factually believing there are apples in it. But when he *sees* an empty drawer, that perceptual experience extinguishes that factual belief. This is an instance of a general point: contrary evidence (other things equal) extinguishes factual beliefs. Imagine this *didn't* happen. Without evidence-based updating, Kevin would *keep going back* to the empty fridge drawer, thinking it had apples. If our factual beliefs didn't update, we'd constantly reperform misguided actions, even when we had the evidence to correct ourselves; this would bode poorly for our survival in a harsh world. But evidential vulnerability doesn't just help explain survival and successful action; it also marks a major difference between factual belief and other cognitive attitudes: one can <u>suppose</u> or <u>imagine</u> contrary to the evidence as much as one likes, but <u>factual beliefs</u> are evidentially vulnerable.

That's a snapshot of the theory. Three points are important going forward. First, any person has a host of factual beliefs that satisfy all four principles (from beliefs about local geography to beliefs about the composition of the paper in a book), and this is because the four principles work together for principled reasons, as we'll see in more detail. Factual beliefs, in the defined sense, *exist* and are widespread. Second and most importantly, *any cognitive attitude that contrasts with factual beliefs on all four principles is not a factual belief—rather, it's a secondary cognitive attitude.* The four principles define the space of cognitive attitudes and mark a great divide in that space. Nevertheless—and this is the third point— even though factual beliefs form a cluster in that space on one side of the divide, some mental states may exist that satisfy some of the principles but not others since the principles are logically independent of one another. A persnickety and somewhat obtuse philosopher might then ask, "Are such states *really* factual beliefs or not?" But this is a nonissue. It will be merely a terminological decision whether we call such mixed states "factual beliefs"; nothing hangs on that terminological decision as long as we specify what we're saying in more detail. The existence of stragglers wouldn't undermine the importance of the principles or the cluster of factual beliefs that they define. My theory is just as useful for understanding the stragglers since it enables us to articulate what's weird about them and formulate hypotheses accordingly.[16] What emerges is a multidimensional property space, where different cognitive attitudes form clusters within that space—"attractor positions," as Sperber or Robert McCauley and Thomas Lawson might put it.[17]

Now for the details. I focus mainly on the contrasts between factual beliefs and imaginings; that's for ease of exposition as well as integration with psychological research. But I also clarify how the contrasts generalize to other cognitive attitudes as well.

3.1. Involuntariness: *If you factually believe it, you can't help believing it.*

It's intuitive that factual belief is involuntary: you couldn't just decide to factually believe that it's 2004 or that you're a billionaire, even if you wanted to. Much ink has been spilled on this issue (voluntarism versus involuntarism), with some philosophers claiming that beliefs *are* under voluntary control. But even philosophers who say beliefs are voluntary still recognize severe limits on the latitude of that voluntariness.[18] Imagining, by way of contrast, has great voluntary latitude. So we can skip past the finer nuances of debates about the voluntariness of belief to make a basic point: voluntary control is a strong differentiator between factual belief and imagining; imaginings are under voluntary control, while factual beliefs are not (or are to a far lesser and severely limited degree).

Here, I develop that basic point by looking at a particular practice of experimental psychologists: *experimenters in psychology often lie to participants*.

Consider, for example, George Quattrone and Amos Tversky's classic experiment in which they attempted to induce (and apparently did induce) self-deception—in particular, self-deception about one's health.[19] Participants held their arms in ice-cold water as long as they could bear, then the experimenters presented fabricated scientific research. They told half the participants that if they had a healthy "Type 1" heart (as opposed to a weak "Type 2" heart), exercise would *increase* their pain tolerance. They told the other half the opposite: if they had a healthy heart, exercise would *decrease* their pain tolerance. The participants then did some distractor tasks before riding an exercise bike and then holding their arms in ice water again. Sure enough, participants in the first group, who were told that a healthy heart made exercise increase pain tolerance, held their arms in the ice water longer than before. But participants in the second group, convinced of the opposite, did the opposite: they held their arms in the ice water for less time than before. But since Quattrone and Tversky had simply fabricated the "research" (there are no such heart "Types," nor does exercise have any effect on pain tolerance), they concluded that the participants were deceiving themselves by changing how long they held their arms in the water.

The important questions for present purposes are these: Why did Quattrone and Tversky *lie* about there being such scientific research? And why is this practice common in psychology?

The practice of lying shows that experimenters *assume* that factual beliefs aren't under voluntary control. Imagine if Quattrone and Tversky had said to participants: "We have a favor we'd like to ask. We'd like you to believe that exercise will increase your pain tolerance if you have a healthy heart. We'll give you additional money if you do decide to believe it for us. We'll give you no evidence; please just believe that for us." Participants would have found the request confusing. They might voluntarily <u>suppose</u>

or <u>imagine</u> they have healthy hearts. They might *pretend* to believe it. But forming an *actual factual belief* just to satisfy a request or make money is not psychologically coherent. If you want someone to factually believe something, your two options are informing them or deceiving them—neither of which brings about a *voluntary* or *chosen* change. Hence, Quattrone and Tversky presented "research" to their participants to make the relevant propositions about hearts and pain *seem like* knowledge and thereby induce belief involuntarily. The same implicit rationale, which assumes involuntariness, carries over to many (if not most) psychological experiments that employ deception.

Many experiments, however, don't require deception. If participants just have to <u>imagine</u> something, no deception is needed. Dan Batson and his colleagues showed that participants were more willing to take a small electric shock instead of letting it be administered to someone else if they imagined the other person's potential pain.[20] Participants who didn't engage in such imagining were less likely to take the shock. But unlike with factual belief, participants were able to imagine at will: they were just asked to imagine what the other person's pain would be, and they chose to imagine it.[21]

The distinction between cognitive attitudes that are involuntary, like factual beliefs, and those that are voluntary, like imaginings, runs so deep that it's a *presupposition* of experiments rather than the conclusion. The methodological assumption is this: *to induce factual belief, either informing or deceiving the participant is required; simply asking nicely or incentivizing are insufficient for factual belief (though those methods work for imagining).* Since this idea lies at the heart of an enormous body of psychological research—in addition to being plausible in light of intuitive examples and one's own phenomenology—we should treat it as a definitional fact about factual belief (in keeping with Quine's Desideratum).

In that light, I propose two definitions of involuntariness. The first ranges over individual mental states; the second ranges over classes. (The word "direct" in each definition is important since one can voluntarily do many things that *indirectly* change one's factual beliefs, like reading a book.)

Involuntariness 1: a mental state x is involuntary if, and only if, (i) x could not have been formed through direct voluntary control on the part

of the person who has it and (ii) x cannot be extinguished or rejected
through direct voluntary control by that person.

Involuntariness 2: a class of mental states X in an individual's mind is
involuntary if, and only if, X cannot be expanded or diminished through
direct voluntary control on the part of that individual.

These give us our first point of contrast in answering Hume's question.
Factual beliefs individually satisfy Involuntariness 1 and as a class satisfy
Involuntariness 2. Imaginings, generally, don't satisfy either; they're vol-
untary, at least to a far greater extent.

This all is mostly straightforward, but let's clear up a couple of issues.

First, some imaginings arise without our choosing them. Kevin might
spontaneously imagine dunking a basketball without having decided to
do so. Some imaginative states are even hard to get rid of, like the arachnophobe's intrusive imagining of the tarantula under the bed. But that
doesn't undermine the utility of involuntariness for distinguishing imaginings and factual beliefs. Imaginings don't satisfy Involuntariness 2: once
one is imagining something, it's easy to choose to dwell on the topic of
the imagining and thereby voluntarily *expand* one's class of imaginative
states. Kevin can choose to imagine dunking a second time, a third time,
and so on, as many times as he likes, making each imagined dunk different. Moreover, closer reflection shows that Involuntariness 1 doesn't apply
to individual imaginings either. Even if one imagines something spontaneously, it remains psychologically possible that one *could have* chosen to
imagine that very thing (as one might do again in the future), which is
not true of factual beliefs.

Second, one might object that self-deception, the very topic of Quattrone and Tversky's experiment, reveals that belief formation can be voluntary: the self-deceived believe what they want (so the objection goes).
But the details of their experiment undermine this objection. The participants deceived themselves *by* holding their arms in ice water for a longer or shorter period. This suggests that the beliefs they self-deceptively
formed weren't just up to them to choose; they had to do something *else* to
convince themselves—in this case, holding their arms in ice water. Self-
deception is not direct voluntary control; it's motivated manipulation (often unconscious) of belief through indirect means.[22]

Involuntariness does help distinguish factual belief from imagining. This contrast applies to other secondary cognitive attitudes as well. When you <u>guess</u> that something is the case, you do so voluntarily: whether you guess that the secret number is 53 or 97 is up to you. When you <u>assume for the sake of argument</u>, you do so voluntarily. And to appeal to Michael Bratman's example, when you <u>accept in the context of budgeting</u> that material costs will be high, you do so voluntarily.[23] And so on. This point is easily checked for any cognitive attitude that is not factual belief. The voluntary/involuntary distinction is thus a deep feature of the space of cognitive attitudes.

We can now answer another question: What does it *mean* to say imaginings are under voluntary control? The answer is that factual beliefs, in combination with desires, form the basic level that structures volition. When Kevin chooses to imagine dunking, he chooses this because he <u>wants</u> to get better at jumping and <u>believes</u> that imagining dunking will help him get better (lots of people imagine in this way in sports practice). In other words, he imagines in a way that would satisfy his wants, if his relevant beliefs (about the benefits of imagining) were true. Thus, beliefs *are partly the basis for choosing what one imagines,* when one does choose. To imagine voluntarily, or to have any other secondary cognitive attitudes voluntarily, is to choose to do so *ultimately* on the basis of desires and beliefs. This point reveals an asymmetric relation: factual beliefs help us choose what to imagine, but that's not true vice versa, since factual beliefs aren't chosen at all. So factual beliefs are cognitive bedrock: they are the unchosen inputs into choice, including our choices of what other cognitive attitudes to form and maintain.

3.2. No Compartmentalization: *Factual beliefs guide actions across the board.*

This next feature, no compartmentalization, highlights another way factual beliefs are cognitive bedrock. No matter what situation you're in, if you make choices, you use factual beliefs to help guide those choices and consequent actions. Factual beliefs, in other words, are not compartmentalized to this or that kind of situation. True, for any given action context, many *particular* factual beliefs won't be relevant and hence won't be involved. For example, your factual beliefs about the planet Saturn don't typically influence how you cook pasta. But that's just a lack of relevance of *particular* beliefs, not compartmentalization of the attitude itself: your factual beliefs about boiling water and pasta, because relevant, will indeed be active. Compartmentalization is thus a qualitatively different *kind* of context sensitivity from mere lack of relevance: relevance activates and deactivates particular mental states of *any* sort, so it is not a differentiator that will help us answer Hume's question about how to differentiate attitude types; compartmentalization, however, applies to secondary cognitive attitudes but *not* to factual beliefs, so it will.

To get clear on this contrast, let's start with imagination and then work back to factual beliefs.

I am imagining a tiny blue elephant on my writing table. That imagining doesn't influence my bodily behavior (except in what I just typed); I'm not acting like the tiny elephant is there, since I'm in everyday work mode (writing). It takes a certain kind of situation, like playing make-believe, to activate imaginings for them to guide behavior. In the practical setting of make-believe, however, I might "hold" or "feed" the imagined elephant by moving my hands in certain ways. People do act on their imaginings in some circumstances. But those imaginings are not in the queue for guiding behavior *across* settings. This is a simple illustration of compartmentalization; you *can* make use of imaginings in guiding behavior, but their influence on bodily movement is limited to certain practical settings (like make-believe, but there are others as well), which you can choose to enter or not.

A passage from Paul Harris's *The Work of the Imagination* substantiates and extends this point. Summarizing experiments with children as young as two, Harris writes:

> Two-year-olds also appreciate the restricted, episodic nature of a pretend stipulation. To show this, we presented two separate episodes one after the other, and stipulated different identities for the same prop within each distinct episode. We then watched to see if 2-year-olds would appropriately tailor their pretence to the stipulation currently in force. . . . For example, we might begin with an episode in which Teddy was said to be having his dinner. Children were handed a brick and asked, "Show me what Teddy does with his sandwich." Children engaged appropriately in pretend feeding with the brick. In a second episode, in which Teddy was

getting ready for bed, children were handed the same brick once more but asked, "Show me what Teddy does with his soap." They now engaged in pretence washing with the brick rather than pretend feeding.[24]

And once the play is over, the brick goes back to being just a brick; the imaginary identities of the brick were compartmentalized to make-believe play. Of course, the children were aware it was a brick all along; it's just that (to use my terms) in the *second* mental map layer (i.e., imagining) they first represented it as a sandwich and then as a bar of soap. To flesh this out, the children have a noncompartmentalized factual belief that the object in question is a brick with certain physical properties; this helps them handle it in a coordinated way whenever they deal with it. But the special setting of make-believe cues the expectation that there should be a second map layer, which can vary by episode and which guides pretense behavior: first, the brick is a sandwich in the second layer; second, it is a bar of soap. Factual beliefs stay active across settings; imaginings (individually *and* as a class) turn on and off for purposes of guiding behavior in and out of the setting of make-believe.

Let's make this a bit more formal. Harris's experiments show that imaginings are compartmentalized on two levels. During the experiments, the kids would have imagined things like this:

IMAGINING 1: *the brick is a sandwich*
IMAGINING 2: *Teddy is eating the sandwich*
IMAGINING 3: *the brick is a bar of soap*
IMAGINING 4: *Teddy is washing with the bar of soap*

The first level is compartmentalization to given *episodes* of pretend play. One episode has imaginary stipulations IMAGINING 1 and 2. The next episode has IMAGINING 3 and 4. As Harris notes, the imaginings from the first episode don't guide behavior in the second; in my terms, each imagining stays in its episode compartment. Still, each episode here is of the *same* practical setting: the setting of make-believe. So the second level of compartmentalization is that imaginings for play turn off for purposes of guiding action outside of make-believe play; for example, when it's time for dinner. (A further point here also helps distinguish compartmentalization from lack of relevance: the content of IMAGINING 1 makes it *relevant* to the dinner setting—it's a sandwich! Yet the child does not eat the brick when it's time for the child's dinner. This illustrates how compartmentalization is a *distinct* psychological feature from relevance; compartmentalization turns a cognitive attitude off, *even when* its contents are relevant, if the practical setting isn't right.)

What exactly is a practical setting? Let's start with the specific setting of make-believe; we can generalize later. That practical setting is made up of three expectations (whether one is alone or in a group). First, someone in make-believe play expects episodes of it to be limited in duration. There will be a start and a stop. Second, people expect certain signals to cue the start and stop. These cues range from subtle (talking in a certain tone of voice) to explicit ("Let's play fire trucks!").[25] The third expectation is that some objects, places, and events will be assigned—in a given episode—values other than what they are factually believed to be.[26] The child knows (hence <u>factually believes</u>) the brick is a brick. But in the make-believe setting, she expects some ordinary objects to be assigned other values, like being a sandwich. Individual episodes of make-believe play differ from one another in that they are different stretches of time and in that participants have signaled different object, place, and event assignments. The point is that accepting this cluster of expectations *both* constitutes entering the practical setting of make-believe *and* (thereby) activates imaginings for purposes of guiding action. To relate this point to the previous subsection, one has voluntary control over what imaginings one has, but one also has a choice of whether to enter the practical setting of make-believe, which then activates one's imaginings for purposes of structuring bodily behavior (like washing Teddy with the "soap").

How exactly does the third expectation work? Kendall Walton talks about the imaginative stipulations and assignments of make-believe play in terms of "props" and "principles of generation."[27] To use Walton's example, a prop might be a tree stump, and a principle of generation might be the assignment that every tree stump counts as a "bear." Then people pretend accordingly, running or preparing to "fight" whenever they come across a tree stump. A principle of generation, then, is a function from props (stumps) to entities one is supposed to imagine (bears) as part of the make-believe game. So being in the setting of make-believe means being prepared to recognize and act on such props and principles for as long as one is in the relevant episode. And once the episode and make-believe setting are over, the imaginative principles of generation and imaginative prop identities are set aside (though often not forgotten, since the game can be taken up again later).

We can now articulate the first important way to see that factual beliefs are not compartmentalized. Pretending a stump is a bear relies on having accurate factual beliefs about what things are stumps (continual reality tracking). If you didn't have a factual belief that there was, say, a stump around the corner, you wouldn't know to imagine a bear around the corner. Since one relies on factual beliefs about props to use those props in pretending, a person's factual beliefs must *not* turn off in the setting of make-believe. In other words, the make-believe compartment does *not* keep factual beliefs out; rather, it relies on them to structure both imagining and behavior since factual beliefs are what keep track of the props on which make-believe action relies.

The structure is this:

FACTUAL BELIEF: *that is a tree stump*
PRINCIPLE OF GENERATION: *IMAGINE tree stumps are bears*
IMAGINING: *that is a bear*

Anything you do on the basis of the imagining (running, saying "that's a bear!" etc.) is therefore *also* done on the basis of the factual belief that something is a tree stump since you need factual beliefs about the props themselves to know what to imagine. This illustrates the crucial difference I've been emphasizing. While imaginings, as a class, turn off for purposes of guiding behavior outside their characteristic practical settings—like make-believe play—factual beliefs stay active *even during make-believe play* in order for that play to work since factual beliefs are crucial for keeping track of the props. Factual beliefs stay "on."

This leads to a second way to see that factual beliefs aren't compartmentalized. We might ask: Why aren't there *any* practical settings in which factual beliefs, as a class, stop guiding behavior altogether? The answer is that factual beliefs also have the job of keeping track of what practical setting one is in. Kevin can join the other kids' make-believe play because he is aware that make-believe play is what's going on, which is to say that he has factual beliefs that represent the practical setting of make-believe. Related factual beliefs also track situational features that get the make-believe started, like that one is on the playground or the stage—or the fact that a friend has said, "Let's play!" Hence, factual beliefs provide conditions for the possibility of pretend play by keeping track of whether one is in the practical setting of make-believe. The factual beliefs that represent that practical setting are ipso facto active *in* the setting. The same point, mutatis mutandis, goes for the practical settings of the other cognitive attitudes, like acceptance in a context. So factual beliefs as a class *can't* be compartmentalized, since compartmentalization itself depends on the situation-tracking work that factual beliefs constantly do: otherwise put, for any compartmentalization, factual beliefs are active in representing the boundaries of the compartment.

There's a third way to see that factual beliefs aren't compartmentalized. Factual beliefs also keep track of the *non*-prop features of the physical environment when one is playing make-believe. Even if the merry-go-round on the playground plays no role in Kevin's pretend play, it's still a physical constraint on his movements. Alternately, the actor onstage—however immersed in her role—must still track the location of the trapdoor so that she won't fall through it by accident. Since make-believe still involves acting in a physical environment, factual beliefs stay active in tracking that physical environment. The same point goes for the practical settings of other secondary cognitive attitudes. To give a philosophical example, let's say I am <u>assuming for the sake of argument</u> that there is no external world as part of a skeptical exercise; still, as I move about the seminar room, I avoid the chair that I factually believe has a busted leg. Factual beliefs help you navigate physically in *any* setting, even one in which you *ostensibly* deny (chairs don't exist!) the very contents of those factual beliefs.

To sum up this subsection so far: (i) imaginings are compartmentalized, while factual beliefs aren't; (ii) factual beliefs make that very feature of imagining *possible* by representing what practical setting one is in; (iii) factual beliefs track prop and non-prop features of one's physical environment regardless of practical setting; and (iv), points (i)–(iii) generalize to the relations between factual belief and other secondary cognitive attitudes (more on that below).

Let's express this portion of the theory more precisely. Factual beliefs individually and as a class satisfy the following two definitions; secondary cognitive attitudes do not ("practical setting independence" is just my technical term for no compartmentalization).

Practical Setting Independence 1: a cognitive attitude x is practical-setting-independent if and only if x guides a person's behavior in all practical settings in which x's content is recognized as relevant to that person's behaviors.

Practical Setting Independence 2: a class of cognitive attitudes X is practical-setting-independent if and only if X is employed in guiding action in all practical settings.

In sum, factual beliefs guide actions across the board, while imagining and other secondary cognitive attitudes do not. Next to involuntariness, this is a second key differentiator that helps with Hume's question.

We can also specify more exactly how factual beliefs *relate* to other cognitive attitudes when it comes to compartmentalization:

Practical Ground Relation: a class of attitudes X is the practical ground of the class Y if and only if individual attitudes in X represent the practical setting one is in such that one acts on representations in Y on account of being in that setting.

As discussed, one acts on <u>imaginings</u> on account of <u>factually believing</u> one is playing make-believe. Hence, <u>factual beliefs</u> satisfy X in this definition when Y is assigned to any of the secondary cognitive attitudes, like <u>imagining</u>. This is a deep part of what makes factual belief the most fundamental cognitive attitude, relative to all the others, which are secondary.

So far, we've mostly focused on factual beliefs versus imaginings—and mostly imaginings for pretend play. Now we can extend these points more completely to other secondary cognitive attitudes, like acceptance in a context, hypothesis, supposition, or assumption for the sake of argument. Let's return to Bratman's cautious budgeter. When budgeting for building materials, she accepts that the material costs will be high (better safe than sorry). Doing this saves her from underbudgeting, which is a far worse error than overbudgeting (better a building that's slightly smaller than ideal than a building that's not complete). But she doesn't act like costs will be high *outside* the context of budgeting—say, when chatting with a friend. Acceptance in a context is also compartmentalized or practical-setting-dependent (as Bratman's name for it, "acceptance in a context," suggests). The practical setting that activates such acceptances is *different* from the setting of make-believe (here it is asymmetry of the costs of errors). But acceptances in a context share with imaginings the simple fact *that* they are compartmentalized and are so in parallel ways: episodes will be of limited duration; they will have starts and stops; and objects, places, and events are assigned values different from what they are factually believed to be. The same point applies to hypotheses and assumptions for the sake of argument. Hypotheses guide behavior in the setting of inquiry. And an assumption for the sake of argument is something you express only in an episode of a certain argument. Once the relevant episodes and practical settings end, hypotheses and assumptions for the sake of argument stop guiding behavior; they, too, are compartmentalized.

The point about factual beliefs' being the practical ground also generalizes. When the building planner acts on her acceptance in a context, it's because she <u>factually believes</u> she's in the setting of budgeting. And so on. It's a general feature of secondary cognitive attitudes that factual beliefs have the power to activate them by representing their characteristic practical settings. This, again, shows another way that factual beliefs are cognitive bedrock: they are the practical ground of the secondary cognitive attitudes.

3.3. Cognitive Governance: *Factual beliefs guide inferences in imagination.*

The picture so far is that factual beliefs are the unchosen (involuntariness) cognitive bedrock that humans use in all settings (no compartmentalization) to keep track of the world and to choose actions in it. We often overlook them even as we rely on them: you don't think of yourself as consulting your <u>belief</u> that your address is 123 Smith Lane; you just go to your home at 123 Smith Lane. Nevertheless, factual beliefs, stored in memory, keep track of your address, so without them, you'd be lost. The bedrock role of factual beliefs is also apparent in more complex actions, like pretending or having a skeptical discussion in philosophy. During pretense, as we just saw, factual beliefs about your surrounding environment (props and non-props) enable you to play make-believe in it. And even if you're in a philosophy seminar skeptically supposing the external world doesn't exist, you still avoid the chair that you factually believe has a broken leg.

This subsection highlights another way that beliefs are cognitive bedrock. Let's extend the example in which your address is 123 Smith Lane. Suppose you're driving with your kids on a long road trip, and the kids start complaining that they want to be home. To alleviate tension, you decide to pretend that you're driving home. "Oh look," you say, "We're almost at our street!" (In fact, you're nowhere near it.) Then one of your kids, playing along, says, "Yay, Smith Lane!" Then you say, "Here's our house." After which the same kid says, "It's 123!"

That example reveals how what we imagine (either for make-believe play or otherwise) uses factual beliefs. Because your kid <u>factually believes</u> your home is on Smith Lane, she imagines approaching Smith Lane when she imagines approaching the street your house is on. Because she believes the address is 123, she imagines arriving at 123 when she imagines arriving home. Roughly put—we'll sharpen this shortly—factual beliefs continuously supply information to the imagination for it to use in elaborating imagined scenarios. As I put it, factual beliefs *cognitively govern* imaginings.

An experimental observation from Paul Harris's book illustrates this point.

> Recall Teddy's bath described earlier. When Teddy was lifted out of the cardboard box he was described as "all wet." Although he had only been bathed in make-believe water from make-believe taps, *the causal powers of make-believe entities are equivalent to those of the real entities that they represent.* So when make-believe taps are turned on they will fill the bathtub with make-believe water; and when something is immersed in that water it gets wet, including Teddy.[28]

Children's background knowledge about real properties guides them from one imagining to the next. This, I maintain, reveals as much about factual beliefs as it does about imaginings (and other secondary cognitive attitudes): factual beliefs supply information for the imagination to use, and imaginings are informed by factual beliefs in this way.[29]

Let's unpack Harris's experiment further. Harris and colleagues would stipulate for the children that something was a faucet. Then they would turn on the "tap." The children would then pretend the "tub" (cardboard box) was filling and Teddy was getting wet. A typical participant would

have had roughly the following sequence of mental states (where there is
in fact no tap, water, or tub).

> IMAGINING A: *the knob on the tap is turned*
> IMAGINING B: *water comes out of the tap*
> IMAGINING C: *the tub fills up*
> IMAGINING D: *Teddy gets wet*

Importantly, each next imagining in the sequence does *not* follow by logic
alone from the one before. IMAGINING B, for example, doesn't follow
from IMAGINING A without background information. Suppose that one
of the children didn't have any idea what a tap was or that water comes out
of one. Then IMAGINING B would not occur to her (if it did, it would be
an unusual stroke of luck or insight). So the following cognitive structure
in most people guides the transition between IMAGININGS A and B:

> FACTUAL BELIEF 1: *turning the knob on a tap typically releases water*

The transition from A to B may *seem* so obvious as to not require any further cognitive support, but it only seems that way to you because you,
dear reader, also have FACTUAL BELIEF 1. Similarly, someone with no
idea what a tub is wouldn't realize it's watertight. So the transition from
IMAGINING B to IMAGINING C is guided by this:

> FACTUAL BELIEF 2: *tubs are large containers that hold water*

Finally, the following factual belief enables the transition from IMAG-
INING C to D.

> FACTUAL BELIEF 3: *water makes things submerged in it wet*

This sequence illustrates my general claim. Cognitive governance means
that factual beliefs (like 1 through 3) are part of the informational background that supports inferential transitions among imaginings (like A
through D).

All this is an elliptical description of what happens in participants'
minds. In fact, far more factual belief structures with information about

the world guide how imaginings unfold: representations of the behavior of liquids, representations of what it is for a container to be sealed, representations of the behaviors of solid objects, and so on. An imaginary stipulation, no matter how vivid, needs further information for a narrative sequence to proceed. And it is part of the functional role of factual beliefs that supply such information.[30] Factual beliefs are not the *only* body of information that influences imaginative transitions. There are genre truths[31] (ideas you don't believe but that provide default assumptions in given fictional genres, like in vampire stories), conventions related to certain types of props, and (of course) what you *choose* to imagine (see above). But, crucially, you wouldn't even *be* imagining a tap unless you drew on your factual beliefs about faucets.[32]

Further research in developmental psychology illustrates how people's factual beliefs fill out their imaginings. Deena Skolnick Weisberg and Joshua Goodstein had child and adult experimental participants fill in unstated elements in a story. They gave participants fictional stories and asked them what else was true "in the story." Most participants maintained that mathematical, scientific, conventional (concerning social norms), and contingent facts were true in the stories, even though the text of the story *didn't state them.* Participants drew on their factual beliefs to fill out what they imagined about the story world. Furthermore, Weisberg, Goodstein, and Paul Bloom show that, contrary to popular myth, children are *more* reality-oriented in their imaginative thought than adults: in their experiments, children were more likely than adults to choose realistic story continuations, even for stories that began in fantastical ways.[33]

So far, you might wonder what the big deal is. After all, *there are many influences on what people imagine, so it is no surprise that factual beliefs enter the fray.*[34] There are, however, two ways in which the role of factual beliefs in guiding imagining is both necessary and fundamental. Here's another way of describing the role in question: when I imagine an object (or some proposition about that object), I thereby also imagine it as having most of the properties (or the most crucial properties) that I factually believe it to have. In terms of a two-map cognitive structure, the point is that the second map layer (imaginings) is constantly importing elements from the first (factual beliefs).

First, factual beliefs have to be playing this role for imaginings to have contents at all. Suppose I tried to imagine a faucet, but at the same time, I tried to *not* imagine the faucet as having any of the properties I factually believe faucets to have: say I imagined a "faucet" as a clump of hardened clay that was perfectly spherical, had no moving parts or openings, and didn't have any water source. Then my imagining wouldn't be of a *faucet* but of something that I merely (inaccurately) labeled "faucet" in my mind. This point generalizes: to imagine anything, my imaginings must be at least partly governed by largely accurate factual beliefs about that very sort of thing. This is true even for imaginings that flagrantly violate reality, like magical fantasies. Even in imaginary fictional worlds like Middle Earth or Hogwarts, the vast majority of objects fall down, make a sound when struck, don't *usually* pass through walls, are invisible when occluded, and so on.[35] The fact that most of these factually believed propositions are tacitly in place is what makes their magical violations in the stories *interesting.* Often, a character has to say a spell for the magic to happen, which shows that the default state of affairs is for things in fantasy fiction to work as they are factually believed to work in reality. So the default filling-in by factual beliefs both makes the exceptions interesting and makes it possible for imaginings to have coherent contents at all.

Second, imaginings don't supply information to factual beliefs *in the way* that factual beliefs supply information to imaginings. That is, the governance relation is *asymmetric* (or, technically, *antisymmetric,* if we're being strict about mathematical terms [see endnote]).[36] This point is easy to misunderstand but is crucial to how factual belief and imagining relate: imaginings do *not* supply the default informational background that governs inferences from one factual belief to the other.

Here's an illustration of this point. Imagine you have in your hand a cup of hot coffee. Now imagine tilting the cup over your other hand. What do you imagine next? Presumably, you imagine the coffee spilling onto your hand and your hand getting burnt. So far, so good: your factual beliefs about the cups, gravity, liquids, heat, and hands are supplying information that fleshes out the imagined scene. But now for the asymmetry. Imaginings *don't* supply information to factual beliefs in this fashion. Suppose that they did. In that case, by imagining I poured hot coffee on my hand, I would also come to <u>factually believe</u> that I had burnt my hand. Why? Well, if imaginings filled in the inferential gaps in beliefs (in the way that beliefs do for imaginings), we would get the following sequence:

FACTUAL BELIEF: *hot coffee burns skin*
IMAGINING: *hot coffee has spilled onto the skin of my hand*
*FACTUAL BELIEF: *the skin on my hand has been burnt*

But do you believe that? No. The factual belief preceded by "*" is the mental state that you don't form (just think: Did you?). You don't infer a <u>belief</u> that your hand has been burnt from merely <u>imagining</u> that hot coffee hit your hand. And that has to be the case: if imaginings governed beliefs, the distinction between factual beliefs and mere imaginings would disappear because all the contents of imaginings would eventually be imported as contents of beliefs. So not only *is* the governance relation between factual beliefs and imaginings asymmetric; it *must* be, on pain of the distinction between the two collapsing. (Sometimes people do come to believe things they imagined, but that is not cognitive governance in the sense I am developing and define below, since it only happens when there is a performance error [like false memory] or when the imagining is realized to comport with other things the agent *already believes,* in which case, it is really just a matter of beliefs governing themselves with imagining playing the role of idea generation [see the endnote for more on this].[37])

I will now define cognitive governance more precisely over classes of cognitive attitudes:[38]

Cognitive Governance: class X of cognitive attitudes cognitively governs class Y if and only if attitudes in X supply the default informational background that supports inferences from elements of Y to new elements of Y.

Factual beliefs satisfy X when imaginings are assigned to Y, but not vice versa. The word "default" here is important because there can be *many* sources of information that support inferences among one's imaginings, like the aforementioned genre truths about vampires that are activated when one reads the relevant genre. The literature on "truth in fiction" is a testament to how complicated such processes can be.[39] But through all that, in the background, are factual beliefs that help us infer what follows from the things that we imagine.

We can now generalize. Factual beliefs also govern the other second-ary cognitive attitudes, like hypothesis, supposition, and so on. Let's say that it's a hot July day and you <u>hypothesize</u> that your cat has been sitting on the furniture. This will lead you to the subhypothesis that there will be cat fur on your furniture, and that's because of your factual beliefs about how cats shed when it's hot. So factual beliefs supply the informa-tion that allows you to infer a subhypothesis from a main hypothesis. Likewise, if you accept in a context that two-by-four pinewood will be three dollars per foot, and you believe that cedar is pricier than pine-wood, then you'll also accept in that context that two-by-four cedar is more than three dollars per foot. Again, you used factual beliefs to infer a further acceptance from your initial acceptance in a context. The same point is true for any of the other secondary attitudes, like supposition or even guessing. If you guess that the ball will bounce higher than a basketball rim, then you also guess that it will bounce higher than ten feet because you factually believe a basketball rim is ten feet high. Any secondary cognitive attitude satisfies Y in the definition of cognitive gov-ernance when the class of factual beliefs is assigned to X. Crucially, it doesn't go the other way.

The overall picture of cognitive governance should be clear. Factual beliefs supply information that enables inferences among all the other attitudes—and among factual beliefs themselves. One further point on how factual beliefs are cognitive bedrock: it's easy to see that every cogni-tive attitude governs *itself—alongside*, as it were, the governance of factual belief. If you imagine that a lion is in the living room and imagine that the living room is a cave, then you'll imagine that the lion is in the cave. Thus, imaginings govern themselves. But just as such imaginings still don't govern factual beliefs (you don't form the *belief* that there is a lion in the "cave"), they also don't govern the *other* cognitive attitudes, like hypothesis, acceptance in a context, and so on. The cognitive governance of factual beliefs, by way of contrast, is *widespread*: factual beliefs have cognitive governance in *all* practical settings and over all other attitudes; other attitudes govern only themselves.

3.4. Evidential Vulnerability: *Factual beliefs respond to evidence.*

Philosophers and psychologists argue over whether humans form beliefs *rationally.* You can pump intuitions in either direction. A person who believes Chicago is north of Houston will be disposed to believe that Houston is south of Chicago; this is both rational and common. But irrational phenomena, like the jealous beliefs of a jilted lover, are common too. Are beliefs characteristically rational or not?

On one side of the debate are thinkers like Daniel Dennett and Donald Davidson who think that rationality partly *constitutes* belief: if a person weren't mostly rational, she wouldn't be capable of believing anything. For Dennett, the very idea of belief is at home in what he calls the intentional stance. Taking the intentional stance involves *assuming* an agent has largely rational beliefs and desires and then predicting what that agent will do on that basis. When you predict that the people around you will walk to the other gate when a gate change is announced at the airport, you're using the intentional stance: you're assuming people will update their beliefs in a rational way and then do the action that would accomplish their goal of flying, given the truth of the updated beliefs. Davidson, inspired by Quine's principle of charity, advocates what he calls "the interpretive view of the mental." On this view, assuming that another being is largely rational is crucial to *interpreting* that being as having beliefs at all. Pockets of irrationality are possible (and common), but complete irrationality isn't. Just imagine someone who said, "North and south are the same kind of arthritis, and so is the number two." What could that person's beliefs even be? A related perspective is that of Fred Dretske, who holds that beliefs need to track objects and properties in the world to have contents at all.[40] Differences aside, all of these views imply that belief formation has to be largely rational. Let's call these thinkers *rationality theorists.*

On the other side are philosophers like Stephen Stich, who argues that Dennett's position is hard to square with psychological facts: humans have irrational tendencies, demonstrated in the lab and in daily life.[41] Mark Johnston argues from the phenomenon of *self-deception* to the conclusion that Davidson's interpretive view is untenable: self-deception, for Johnston, produces irrational beliefs, so rationality doesn't constitute belief.

More recently, Lisa Bortolotti takes aim at Davidson's view by appealing to psychiatric delusions: a delusion like Capgras syndrome, whereby a person claims a partner or family member is an imposter, is one example among many of irrational beliefs, and such irrationalities are widespread enough to undermine Davidson's view. And Eric Mandelbaum and Jake Quilty-Dunn argue from various biases that human belief formation is not rational one way or another. Let's call these thinkers *irrationality theorists*.[42]

Rationality theorists can respond by saying two things. First, irrationalities are comparatively rare against a typical person's large background of rational beliefs. Even someone with Capgras syndrome usually knows what his address is, what toaster ovens are, what faucets do, and so on for myriad everyday beliefs. Second, many of the irrational "beliefs" that the irrationality theorists appeal to are only "beliefs" in an attenuated sense—or are not beliefs at all. Tamar Gendler contends that self-deception produces not beliefs but a *pretense* mental state, something like wishful imagining. Jason D'Cruz argues similarly that rationalization (for example, "It's not cheating because everyone else is doing it too") is "performative pretense."[43] This second response, however, points to a methodological quandary: Should the existence of irrational beliefs lead us to reject the rationality theorists' theories (as Bortolotti argues), or should those theories lead us to say that such irrational "beliefs" aren't *really* beliefs (as Gendler argues)? The quandary is that it's not obvious where to start.

The solution is to recognize that beliefs are real phenomena in the minds of real people; our job is to theorize about *those* mental states and describe them as accurately as possible. Typical humans do have impressively rational factual belief sets *when it comes to everyday life* (that is, when we set aside ideological or identity-related "beliefs"). So, among other things, we need to describe *those*. My running example of beliefs about how to get home illustrates this: a typical person has dozens of rational beliefs about her local neighborhood (what buildings are where, which street is parallel to which, where the train station is, where the stores are, etc.). And consider, to pick an item at random, the beliefs you have about automated teller machines (ATMs). You probably believe that *ATMs have buttons, ATMs have screens, ATMs operate on electricity, ATMs store money, ATMs take bank cards, ATMs only give money if you enter your pin, ATMs charge fees when they're not from your bank, ATMs distribute bills and not change, ATMs in other countries distribute the currencies of those countries, your bank charges a fee when you use an ATM in another country,* and so on. Sound familiar? That's because you already have those rationally formed beliefs. And an ATM is just *one* kind of object out of thousands you know much about. So humans have rational belief-formation mechanisms. But it's easy to ignore them—as I think irrationality theorists too often do— since they usually operate swiftly and efficiently enough to escape notice.[44] So we should side with rationality theorists on two points: (i) rational belief-formation mechanisms *exist* in normal humans; and (ii) normal humans have a broad background of rationally formed factual beliefs that allow navigation and action in the world.[45]

We saw above, furthermore, that we have voluntary control over our imaginings, but we can't form beliefs at will. This raises the question of what constrains beliefs but not imaginings. A candidate answer is that beliefs are more constrained by rational coherence with evidence than imaginings (or other cognitive attitudes).[46] You can voluntarily *imagine,* for example, all kinds of ATMs—even sci-fi ones that scan your eyes to confirm identity—but your *beliefs* about ATMs are constrained by the evidence you've encountered (those you've seen, heard about from reliable sources, etc.).

So some rational constraint separates factual beliefs from other cognitive attitudes. But it can't be too strong (and it won't be the *only* influence on belief formation), since the psychological evidence for biases is extensive.

Accordingly, I propose the following definitions of evidential vulnerability, which is a kind of rational constraint. These definitions will help us sail between the Scylla and Charybdis of positing too much or too little rationality—*and* it will help answer Hume's question of how believing and imagining differ.

Evidential Vulnerability 1:

(i) If cognitive attitude x is involuntarily prone to being extinguished if (a) it conflicts with perceptual states or if (b) it is realized to lead to a contradiction, then x is evidentially vulnerable.

(ii) If cognitive attitude x is involuntarily prone to being extinguished if it contradicts or does not cohere with other evidentially vulnerable states, then x is evidentially vulnerable.

(iii) No other cognitive attitudes are evidentially vulnerable.

This definition is recursive, meaning that the second clause extends the class captured by the first (and so on). That's because when one mental state is vulnerable to evidence, another mental state can be vulnerable to evidence by being vulnerable to the first one, and the recursive clause (ii) captures this.

I define evidential vulnerability for classes of mental states as follows:

Evidential Vulnerability 2: a class of cognitive attitudes X is evidentially vulnerable if and only if X is composed only of attitudes that are evidentially vulnerable as defined in Evidential Vulnerability 1.

Factual beliefs satisfy these definitions, while secondary cognitive attitudes do not. Before arguing for that position in detail, however, let me explain how the definitions work.

The idea is to capture the rationality of factual beliefs by focusing on their *extinction conditions*: what makes mental states go away. If you thought it was raining and look out the window and see that it isn't, then—*poof!*—your previous belief goes away. That's because visual input is evidence and factual beliefs are vulnerable to evidence. Of course, many things make factual beliefs go away—such as simple forgetting. But positing evidential vulnerability as *a* major extinction factor for factual beliefs threads two important needles. First, it identifies a rational aspect of factual beliefs *without* denying there are also irrational influences. Second, it identifies a rational feature that distinguishes factual beliefs from other cognitive attitudes while still allowing that other attitudes can also be responsive to reason in other ways. If we look at the history of scientific thought, we see that there have been *many* rational suppositions and hypotheses. How then do they differ from factual beliefs? The answer is that one can *voluntarily* maintain them in the face of defeating counterevidence: one can suppose or hypothesize ideas that one already knows are wrong; there may be educational value in doing so. So those attitudes, unlike factual beliefs, don't satisfy the specific definitions just given: in particular, they lack the evidential extinction conditions of factual belief.[47]

Now here's a theoretical argument for why factual beliefs are evidentially vulnerable. Evidential vulnerability supports factual beliefs in guiding actions that satisfy desires. Consider again the example of Kevin wanting an apple. If he thinks there are apples in the cupboard, he goes to the cupboard. If he thinks they are in the fridge, he goes to the fridge. His factual beliefs guide his behavior to get him what he wants. But now suppose his factual beliefs were *not* vulnerable to evidence. Then, even if he *saw* no apples in the fridge upon looking, his factual belief that they were in the fridge would remain since, lacking evidential vulnerability, they would *not* extinguish in light of the contrary visual evidence. So what would he do? He'd go right back to the fridge again and again (!) since his factual belief that there are apples in the fridge would not be extinguished. We can invent absurd scenarios like this ad nauseam. The evidential vulnerability of factual beliefs is crucial to our ability to satisfy our desires since, without it, we'd constantly reperform actions that are useless. Furthermore, it is by trying to satisfy desires that we often go into situations that provide evidence contrary to our factual beliefs, such that we update them: Kevin's desire for an apple led him to the very situation that forced him to update his beliefs about where the apples were. So there are two tight connections between evidential vulnerability and the action-guiding role of factual beliefs: (1) evidential vulnerability supports desire satisfaction, and (2) seeking to satisfy desires often leads us to situations that force factual beliefs to update, given their evidential vulnerability.

There are, furthermore, many empirical reasons in favor of the evidential vulnerability of factual beliefs. I cover just a few here, drawing again from developmental psychology.

Alison Gopnik has long advocated the famous "theory theory" of how children learn about the world:

> The basic idea is that children develop their everyday knowledge of the world by using the same cognitive devices that adults use in science. In particular, children develop abstract, coherent systems of entities and rules, particularly causal entities and rules. That is, they develop theories. The theories enable children to make predictions about new evidence, to interpret evidence, and to explain evidence. Children actively experiment with and explore the world, testing the predictions of the theory and gathering relevant evidence. Some counter-evidence to the theory is simply reinterpreted in terms of the theory. Eventually, however, when many predictions of the theory are falsified, the child begins to seek alternative theories. If the alternative theory does a better job of predicting and explaining the evidence it replaces the existing theory.[48]

Gopnik points out that researchers have applied this account to children's understanding of the physical world, the biological world, and the psychological world (children's understanding of minds). In each domain, infants are born with basic initial innate theories or concepts, which they revise, reject, or extend in light of incoming evidence. This is why play is important for more than just fun: it provides children with experiences that confront their theories and lead to new ones. None of this is to be interpreted in an overly intellectualized way: children don't conceive of what they do as "theory revision"—to them they're just playing. It's just that young minds are so structured that their beliefs about things in the world—their "theories"—update in much the same ways that scientists deliberately update their theories.

Renée Baillargeon's research on infant learning and cognition is also relevant, especially her research on how infants learn *object statics* (this refers to physical facts about how objects stack, what can fit inside what, etc.). We humans learn a great deal of object statics in the first year of life. Infant humans come to understand the causal structure of object stacking much better than do chimpanzees, our close primate relatives.[49] Summarizing a series of experiments with realistic and unrealistic combinations of stacked boxes, Baillargeon writes:

> By 3 months of age, infants have formed an initial concept of support centered on a simple *contact/no-contact* distinction: they expect the box to remain stable if released in contact with the platform, and to fall otherwise. At this stage, any contact with the platform is deemed sufficient to ensure the box's stability. In the months that follow, infants identify a sequence of variables that progressively revise and elaborate their initial concept. At about 4.5 to 5.5 months of age . . . infants begin to take into account the *type of contact* between the box and the platform. Infants now expect the box to remain stable when released on but not against the platform. At about 6.5 months of age, infants begin to consider the *amount of contact* between the box and the platform. Infants now expect the box to remain stable if a large but not small portion of its bottom surface rests on the platform. Finally, at about 12.5 months of age, infants begin to attend to the *proportional distribution* of the box; they realize that an asymmetrical box can be stable only if the proportion of the box that rests on the platform is greater than that off the platform.[50]

This impressive progression shows how it is a cognitive accomplishment to learn basic features of the world that we adults take for granted. More interesting still is Baillargeon's account of *how* that learning takes place. Her view is that infants start out with a few core principles and concepts, including *solidity, continuity,* and *force.*[51] But they also have a learning mechanism that tracks "contrastive outcomes" that allows them to identify relevant variables and thus build these core principles into more sophisticated rules.

Thus, on Baillargeon's theory, infants have

1. A small number of innate principles and concepts.
2. Contrastive outcome learning: the ability to form rules out of the innate principles plus observed variations in outcomes by identifying difference-making variables.
3. The disposition to play in ways that provide contrastive experiences (e.g., stacking blocks, etc.).

From these few elements, an impressive understanding of object statics emerges in an infant's first year of life.

The infant object statics that Baillargeon describes is, as I see it, a special case of Gopnik's theory theory. In this case, the rules that infants form are a "theory" they have about physical objects, and the contrastive outcomes by which they revise their rules are their evidence. All of this is an extended illustration of evidential vulnerability. In the course of learning, infants form factual beliefs—basic representations of how the world is for them—and it is clear that they discard or revise their beliefs in light of the evidence constituted by incoming experiences.

Evidential vulnerability doesn't appear only in how infants or young children learn from firsthand experience. It also appears in how young children learn from reports from other agents, as Paul Harris and colleagues show in experiments on how young children discriminate between different testimonial sources.[52]

The experiments in question, which have many permutations, start with two different characters (for example, a Rhino figure and a Lion figure) who give true or false reports about objects already familiar to the child participants. Rhino might call a pencil a "fork," while Lion calls the pencil a "pencil." Next, the experimenters observe which of the characters the children trust about items that are *unfamiliar*. It turns out children discriminate between agents who have said known-to-be-true things in the past and those who have said things they knew to be false, and they preferentially rely on the former for learning *new* things. So they'll believe Lion when he labels a new object a "wug," but they're much less likely to believe Rhino.

Sunae Kim, Charles Kalish, and Paul Harris show that children between three and five also track speaker reliability to figure out which categories to use in inductive inference—in figuring out what *else* is likely to be true about new objects they encounter.[53] In the test phase of these experiments, Lion or Rhino would apply labels that grouped together unfamiliar objects that looked different from one another; for example, two dissimilar-looking objects might both be labeled "dax," while an object that looked similar to one of the first two might be labeled "wug." Children then faced the choice of doing inductive inference on the basis of *appearances* or on the basis of the *labels* they had heard from Lion or Rhino.[54] Sure enough, when a speaker who had been reliable in the past applied the same label to dissimilar-looking objects, children would use the labeled category rather than the appearances for doing inductive inference. But when a past unreliable speaker labeled dissimilar-looking objects under the same heading, children ignored the label and relied on appearances.

These testimonial trust experiments show (i) that a child's beliefs about a speaker's reliability will be revised in light of evidence and (ii) that beliefs about speaker reliability, which are themselves evidentially vulnerable, allow for the generation of new factual beliefs that can be used in downstream thought. *Pace* the irrationality theorists, all of that is rational, and it exemplifies evidential vulnerability in particular. So we can conclude this section with the following point: at least one class of "beliefs"—which I call factual beliefs—is generally rational in at least one specific way: factual beliefs are evidentially vulnerable.[55]

3.5. Factual Beliefs and Secondary Cognitive Attitudes: *The big picture.*

We can now cash out the distinction that motivated this chapter. Recall that we wanted to know the difference between these two mental states.

1) Sam <u>factually believes</u> *that p.*
2) Sam <u>fictionally imagines</u> *that p.*

Wherein consists the difference? asks Hume. I respond that the first means that Sam has a cognitive attitude toward *p* that (in the defined senses) is involuntary, is practical-setting-independent in its action guidance, has widespread cognitive governance, and is evidentially vulnerable. The second means that Sam has a cognitive attitude toward *p* that (in the defined senses) is voluntary, guides action specifically in make-believe play, has only limited cognitive governance, and is not evidentially vulnerable. That's how I answer Hume's question.

*   *   *

Even more abstractly, what is my overall portrait of the space of cognitive attitudes?

Let's start with a correction and then build from there. Philosophers have a slogan about belief: "beliefs are the map by which we steer the ship."[56] But that's misleading: sometimes, as we've seen, we steer the ship with representational states that we don't actually (factually) believe: hypotheses, assumptions for the sake of argument, suppositions, fictional imaginings, acceptances in a context, and so on all can and do guide our actions. So the slogan is too crude.

Nevertheless, factual beliefs are still fundamental.

We generate instances of the other attitudes when we choose to, and we use factual beliefs to help *decide* when to engage in such imagining (or hypothesizing, etc.).

We use secondary cognitive attitudes in guiding our actions, but we do that when we *factually believe* we are in the appropriate practical settings: settings in which we choose to rely on what Bratman calls "an adjusted cognitive background" and I call the second layer in a two-map cognitive structure.

When we use secondary cognitive attitudes in thought or action, our factual beliefs supply information to help guide the inferences we draw (cognitive governance).

Finally, factual beliefs remain involuntarily tethered to the world through evidential vulnerability, directly or indirectly, in a way that keeps them largely accurate and thereby helps us avoid ditches and avoid per-petually reperforming the same misguided action.[57]

The overall picture is that factual beliefs are tightly tethered to the world through evidential vulnerability, but the secondary cognitive at-titudes are—granting latitude for voluntary choice—tethered to factual beliefs through cognitive governance, practical setting dependence, and practical grounding.

Let's correct the slogan. Factual beliefs are not the only map by which we steer the ship. But factual beliefs are the basis on which we choose, extend, and evaluate the other cognitive attitudes: factual beliefs are thus *conditions for the possibility* of having and using the other maps—the other cognitive attitudes—by which we sometimes steer.

4. DESIDERATA SATISFIED/HANDOFF TO RELIGIOUS CREDENCE

How does this theory fare with respect to Hume's Desideratum, Quine's Desideratum, and Clifford's Desideratum?

Hume's Desideratum says that a theory of belief should distinguish it from imagining and other cognitive attitudes. How mine does this is by now clear.

Quine's Desideratum says that a theory of belief and other cognitive at-titudes should unify philosophy and empirical psychology on those mat-ters. My theory does this in two ways. First, I elaborated it by appealing to extensive research in psychology, ranging from the work of Quattrone and Tversky, to the work of Harris, to the research of Gopnik and Bail-largeon, and others. Second, my theory helps explain *other* empirically demonstrated features of action; in particular, pretense action. Harking back to the Prologue, let's recall that make-believe play has three distinc-tive features: a *two-map cognitive structure, nonconfusion,* and *continual reality tracking.* Here's how my theory explains them. By differentiating factual belief from other attitudes, I explain how the two map layers in pretend play are distinct. Nonconfusion is explained by the asymmetry of cognitive governance: since factual beliefs supply information to guide inferences among imaginings but not vice versa, the factual beliefs are effectively quarantined (to use a term from Gendler) from imaginings;

this gives us nonconfusion. Finally, continual reality tracking is explained by evidential vulnerability. Since those aspects of pretending are empirically validated, a theory (like mine) that explains them satisfies Quine's Desideratum.

Clifford's Desideratum tells us to explain why we hold beliefs (or at least factual beliefs—more on this in Chapter 5) to more stringent norms than other attitudes. The answer my theory gives is that the effects factual beliefs have in a person's cognitive economy are far more pervasive than those of any of the other cognitive attitudes. Since factual beliefs represent basic features of the world across situations, their falsity is likely to cause problems in *any* situation—leading us to do actions that frustrate us or get us to fall into a ditch. Insofar as we care about our own or others' success, it makes sense for us to worry whether they're being formed rightly. Furthermore, since factual beliefs are involuntary, one cannot change them in someone else simply by offering a reward. One must appeal to the sorts of levers to which factual beliefs are vulnerable; namely, evidence—or at least what is taken for such. On this view, norms of truth and rationality for factual beliefs come out as guidelines for appealing in oneself or others to the sorts of inputs that factual beliefs are responsive to anyway, and the reason people think one *should* appeal to such norms is that falsity in one's factual beliefs leads to failures of action across the board or to actions in others that would frustrate one's own goals. On the flip side, imaginings and other secondary attitudes can be false with no or little negative effect on successful action, and that is so for two reasons: first, they are limited to their own practical settings; second, even when they are active in guiding behavior, factual beliefs are still active in the background of the two-map cognitive structure of which imaginings are the second layer—keeping us on course in even our most fanciful moments. My view is that humans sense these differences—however vaguely—which is partly why we have words that express different attitudes and why we apply differential norms to them.

In the next chapter, I put this framework to work. I show that many people's religious "beliefs" also lack the defining features of factual beliefs. Conversely, they share defining features with the other secondary cognitive attitudes, like fictional imaginings. Such religious credences, even though they often have contents that purport to describe how the world is, are not factual beliefs.[58]

CHAPTER THREE

Religious Credence Is Not Factual Belief

1. TWO CAMPS ON RELIGIOUS "BELIEF"

The work of Chapters 1 and 2 enables us now to address the crucial question clearly: *What <u>cognitive attitude</u> or <u>attitudes</u> do religious "believers" have toward their stories and doctrines about deities, demons, angels, dead ancestors, and other supernatural entities?*

Chapter 1 zeroed in on the attitude dimension of people's psychological states. Chapter 2 gave a theory of cognitive attitudes. That theory now provides a stable framework for classifying religious cognitive attitudes, given suitable evidence, thereby answering the question just posed.

But before we get there, we need some framing.

When it comes to our question, two types of views have emerged: One-Map Theories and Two-Map Theories (of the sort I advocate). Importantly, there are *general* and *particular* varieties of each.

Let's start with a particular One-Map Theory. Concerning the Fang people of Central Africa, Pascal Boyer writes:

> The Fang . . . are exposed to a whole range of supernatural objects, beings and occurrences that are taken *in a very matter-of-fact way* as part of daily existence. Witches may be performing secret rituals to get better crops

than you. Ghosts may push you as you walk in the forest so that you trip and get hurt. Some people in the village are presumed to have an extra organ and may prove to be very dangerous. Indeed some people are widely believed to have killed other people by witchcraft, although nothing could be proved. I do not mean to suggest that Fang people live in a paranoid world with ghouls and monsters lurking in every corner. Rather, ghosts and spirits and witchcraft are part of their circumstances *in the same way as* car crashes, industrial pollution, cancer and common muggings are part of most Western people's.[1]

To put the point into my terms, Boyer portrays Fang individuals as factually believing such ideas as *witches perform secret rituals, ghosts push you in the forest, some people have an additional organ with dangerous powers,* and *people sometimes get murdered by witchcraft.*[2] His phrases "matter-of-fact way" and "in the same way as" (followed by ordinary events) gesture at the same factual belief attitude described in the last chapter. On this description, the Fang have a *one-map* (factual belief) cognitive structure that in general represents *both* propositions about straightforwardly natural entities, like trees and people, *and* propositions about supernatural entities, like ghosts and witchcraft.

The central claim of any particular One-Map Theory thus has this logical shape:

> **Particular One-Map Theory**: people <u>exist</u> (in such-and-such a community) who mostly factually believe their religious and other supernatural ideas.

Different particular One-Map Theories might also say that factually believing supernatural ideas is *typical* for people in a given community, like the Fang. But they do *not* go on to claim this is how things generally work in people's minds around the world.

But some thinkers would claim that something like Boyer's view about the Fang holds for religious "believers" in general. That gives us a *general* One-Map Theory: religious people *generally* have the same attitude toward the existence of their deities as you or I have toward the existence of electricity or household furniture; they just think (factually believe) they exist. Maarten Boudry and Jerry Coyne advocate a general One-Map Theory:

There are numerous examples of religiously motivated behavior that make perfect sense given the assumption that people *actually and factually* believe them, and that make little sense on the supposition that they don't . . . religious beliefs are just like ordinary beliefs . . . some people really do believe in a 6-day creation or in 72 virgins awaiting them in paradise. . . . By and large, "religions" consist of factual claims about the nature of the universe, endorsed and acted upon by millions.[3]

Like Porot and Mandelbaum, Boudry and Coyne conflate attitude and content. Contrary to how they reason, one can have all sorts of *different* attitudes toward any given "factual claim" (supposing, hypothesizing, etc.), so the presence of such claims in religions is not decisive as to whether people factually believe those claims in the relevant <u>attitude</u> sense. Still, what they are trying to conclude is clear. For all "genuine" religious persons, their factual belief maps include the entities described in their religious stories and doctrines: one layer of cognitive processing for natural and supernatural ideas alike.[4]

The central claim of any general One-Map Theory thus has this logical shape:

> **General One-Map Theory**: <u>all</u> (or almost <u>all</u>) religious people factually believe their religious and other supernatural ideas.[5]

This is the view you get if you stitch together particular One-Map Theories about *all* or almost all religious people in the world.[6]

Two-Map Theories, on the other hand, posit in people's minds two layers of processing. On my Two-Map Theory, a given religious person has a factual belief layer that mostly represents ordinary stuff and a religious credence layer that's a secondary cognitive attitude.

But Two-Map Theories also come in particular and general varieties. A particular Two-Map Theory has this logical shape:

> **Particular Two-Map Theory**: people <u>exist</u> (in such-and-such a community) who have a distinct cognitive attitude of religious credence.

A general Two-Map Theory has this logical shape:

> **General Two-Map Theory**: <u>all</u> (or almost <u>all</u>) religious people have a distinct attitude of religious credence toward their supernatural or religious ideas.

Now here's the important part. The general theories in the different camps contradict one another (obviously). And a general theory in one camp is at odds with any given particular theory in the other. *However,* most particular One-Map Theories do *not* contradict most particular Two-Map Theories. One could easily, for example, hold a particular One-Map Theory about the Fang and a particular Two-Map Theory about, say, the *pujari* who assist worship in Hindu temples in India. On that combined view, the Fang would factually believe there are supernatural entities like witches, while the *pujari* would have religious credences (not factual beliefs) concerning their supernatural doctrines and stories. It is an open question, of course, whether such a view is true; we shouldn't judge either way before considering relevant evidence. Humans are complex, capable of great cognitive flexibility and cultural variation. So we shouldn't be surprised if different groups of people or even individuals held different attitudes toward their respective religious and other supernatural ideas.

My stance is this: *many* (and probably most) people around the world have two-map cognitive structures for processing their religious ideas: a factual belief layer and a religious credence layer. But empirical exploration is required when it comes to any particular religious community to work out what attitude(s) people in that community have toward their stories and doctrines. Neither a Two-Map Theory nor a One-Map Theory should be the default stance; rather, we should adopt whichever *particular* theory best explains the relevant data and then expand our explanatory scope from there as our evidence base grows. So my goals in this chapter are two: (1) to show how to apply the theoretical tools introduced in the last chapter to cases of religious "belief" and (2) in so doing, to show that there exists a secondary cognitive attitude of religious credence. The way to accomplish both goals is to start by arguing for a *particular* Two-Map Theory in relation to a religious community about which we have substantial data. In the next chapter, we'll expand our evidence base around the world and back in time as a way of showing that my approach has broader application as well.

The religious community I focus on here is the Vineyard Movement, an evangelical, neo-Pentecostal Protestant denomination with around twenty-four hundred congregations in North America, Europe, Australia, New Zealand, and South Africa. Like several other evangelical

Christian denominations, the Vineyard emerged in Southern California in the 1970s and expanded from there. It is also representative of much of American evangelicalism insofar as it emphasizes a personal relationship with Jesus over strict doctrine and has a modernized, charismatic style of worship (think: guitar and PowerPoint instead of organ and hymnal). I choose this focus because two ethnographies have recently appeared that give us extensive anthropological data on the Vineyard: Tanya Luhrmann's *When God Talks Back* and Jon Bialecki's *A Diagram for Fire*.[7] Those data show the Vineyard to be an excellent case study for exploring religious cognitive attitudes. Consider this passage from *A Diagram for Fire*:

> As we will see, naturalistic double coding of supernatural phenomena is a common framing in the Vineyard. Like a shady legal operation that has *two different sets of books*, parallel naturalistic and supernaturalistic accounts are often produced concurrently about the same phenomenon.[8]

We will return to this "double coding" repeatedly in what follows. Protestant Christianity, moreover, is widely studied in the psychology of religion, so ethnographic data on the Vineyard can be usefully compared to controlled psychological research on the broader category. Taking these points together, I argue that the converging evidence is best explained by a particular Two-Map Theory of Vineyard "belief": that is, many Vineyard "beliefs" in the heads of individual members are religious credences and not factual beliefs, and they relate in interesting ways *to* their factual beliefs.

Note that my view does *not* entail that members of the Vineyard lack fervor, since fervor itself doesn't necessarily imply factual belief. The list of charismata professed in the Vineyard is impressive: speaking in tongues, being slain in the spirit, miracle healing, prophecy, hearing God's voice, deliverance from demons, lying on graves, and communication from God through encounters with randomly selected biblical texts. A One-Map Theorist may look at that list and say, "Surely they simply, factually believe there are such things as God, demons, miracles, and so on!" Yet the data, I argue, show otherwise: Vineyard members oscillate between a map of the world in which such things exist and a map of the world in which they don't.

2. THE SECOND MAP LAYER: RELIGIOUS CREDENCE

Let's start with an observation from Erving Goffman's *The Presentation of the Self in Everyday Life* ("teams" here refers to social groups):

> The performance given by a team is not a spontaneous, immediate response to the situation, absorbing all of the team's energies and constituting their sole social reality; the performance is something the team members can stand back from, back far enough to imagine or play out simultaneously other kinds of performances attesting to other realities. Whether the performers feel their official offering is the "realest" reality or not, *they will give surreptitious expression to multiple versions of reality, each version tending to be incompatible with the others.*[9]

As an example of "surreptitious expression," think of an actor who adjusts her head microphone when the audience is looking elsewhere. This action reveals that at a basic level, she still represents herself as being *merely* an actor and not her imagined character (why would Hamlet's mother even have a head microphone?). Similarly, various things religious people do reveal that at some level they represent the world as being *merely* a natural world, even if they say otherwise. Consider something subtle that Bialecki observed when attending a "living room seminar" of one Vineyard church. In this situation, a widely known Vineyard "prophet" (that's what the members of the group called him) was visiting the small group, and in the prayer session before his address, he was speaking in tongues.

> When all the chairs in the living room were filled and people had started to sit on the floor, one of the coleaders picked up the guitar, prayed "come, Holy Spirit," and started playing plaintively and slowly the songs that were familiar from church services and worship music CDs and downloaded MP3s. The prophet covered his face with his hands as he started rocking back and forth in time to the music. In the background, the rustling whisper of the "polite," subvocal speaking in tongues could be heard. After a while, the prophet joined in speaking in tongues with his eyes closed, though at times he would open them, stop speaking in tongues, and check his watch—as a sign not of bad faith but of nonchalance.[10]

So a man known for his spiritual charisma would be so infused with the Holy Spirit that he spoke in tongues, yet he would also break his speaking in tongues to check his watch.

There are many ways to interpret this event. But there is a striking similarity between (a) the actor onstage who, though in character, adjusts her head microphone and (b) the prophet who, though in character (speaking in tongues), checks his watch. In both cases, the acting agent has a dramatized mental map of what's going on, in which she or he is swept up in an epic or spiritual arc of events, and a mundane map of what's going on, in which head microphones are falling off or the long evening is threatening later plans. The microphone-adjusting and the watch-checking surreptitiously express the mundane, factual belief map.

Importantly, that is not to say that the watch-checking prophet was necessarily consciously *aware* of the resemblance of his cognition and behavior to that of a performer. He may or may not have been, which is worth keeping in mind: part of the point of Goffman's work is to show that such resemblances to theater pervade "everyday life" (social settings, behaviors in restaurants, etc.), *whether or not we are aware of those resemblances.* So for one who takes Goffman seriously, it is no objection to my position to say that the Vineyard prophet may not have consciously thematized his actions as make-believe play: people pretend all the time without consciously thematizing it as such. In any case, as Luhrmann repeatedly points out, many Vineyard members *do* consciously characterize much of their religious behavior as pretending (as we'll see below). So the interesting open question (which I take up in Chapter 5) of how much second-order awareness people have of the difference between their religious credences and their factual beliefs can—and does—admit various answers in various cases that are all compatible with the first-order distinction itself.[11]

The watch-checking *could* be interpreted (awkwardly) in light of a One-Map Theory. The claims at this point are high-level enough that sufficient tinkering can make either theory noncontradictory with any given fragment of the data, even if the fit is ill. But as evidence builds up and we see more apparent surreptitious expressions of two distinct cognitive maps, the One-Map interpretation becomes less tenable. And any evidence of "double coding," as Bialecki puts it, also supports a Two-Map Theory.

My theory of cognitive attitudes from the last chapter is suited to provide the relevant Two-Map Theory and thereby make sense of both double coding *and* surreptitious expression, or so I seek to demonstrate below. Recall the four principles of factual belief.

1. *If you factually believe it, you can't help believing it.*
2. *Factual beliefs guide action across the board.*
3. *Factual beliefs guide inferences in imagination.*
4. *Factual beliefs respond to evidence.*

As we've seen, these principles differentiate factual beliefs (to which they apply) from secondary cognitive attitudes (to which they do not). I apply them here to Vineyard data and thus show that many Vineyard members' religious "beliefs" differ from factual beliefs in all four ways: they are held voluntarily, are practical-setting-dependent (compartmentalized), lack widespread cognitive governance, and don't respond to evidence. So these Vineyard "beliefs" are secondary cognitive attitudes and hence form a second layer of cognitive processing of ideas. We should therefore regard them as the distinct cognitive attitude of religious credence.

2.1. The Voluntariness of Vineyard "Beliefs"

Many Vineyard members "hear" or attempt to "hear" the voice of God. (I put "hear" in scare quotes because Luhrmann and Bialecki both make clear that the "hearing" is almost always in the head—auditory mental imagery rather than perception or hallucination.) In fact, many of them practice for hours each week cultivating their auditory imagery so that it *feels to them* like it comes from an outside source. Cultivation of imagery ranges from sensory deprivation (sitting in a silent room) to games of pretend play, like setting an extra place for God at the dinner table. A message that arrives from outside-feeling auditory imagery *may* be regarded as the voice of God *if* it conforms to certain rough rules.

Say a Vineyard member has auditory imagery of a voice telling her to go on a mission to Mexico.[12] When would that imagery be counted as God's voice? According to Luhrmann, four conditions typically have to be satisfied. First, the imagery's content has to be at least somewhat *surprising*—not something one would usually come up with on her own.

Second, it has to be something that might well come from God—it has to seem characteristic of (one's conception of) Him. Third, it should comport with other circumstances or other people's prayers—the message of the auditory imagery should relate interestingly to *some* independent event. Fourth, the "hearing" should give "the feeling of peace."[13]

All of these conditions are open-ended, however. They rule out many things, but they leave a lot to be settled by discussion and debate with oneself or others (*was the voice saying to go on a mission "really" God's?* etc.). And this latitude is one place where the voluntariness of "beliefs" about "hearing" God becomes apparent. One of Luhrmann's informants puts it like this:

> I can *choose* to believe this [auditory mental imagery of a voice] is from God, or I can think this is just from me, and the reality is that it could be either, and I know that. There is always a *choice* to believe what it is.[14]

Luhrmann relates this ability to choose to "believe" to the Vineyard members' awareness that the internal ideas and images they label as "from God" are (quite probably) from inside themselves. The same informant adds this: "Sometimes when we think it's the spirit moving, it's just our burrito from lunch."[15] Relatedly, Bialecki notes that it's common inside the Vineyard to joke about the difficulty of determining whether the feelings they're having are from God or from the pizza they had for lunch. The result of this underdetermination of how to interpret images that come to mind—so common that it's a running joke—is that, even if a certain image appears as a surprise, forming the "belief" that it's from God is still a matter of choice. Vineyard members say as much openly. There is also a voluntary step earlier in the process of Vineyard "belief" formation when it comes to "hearing" God. Luhrmann writes:

> They learn to infuse the absent, invisible being with presence by cherry-picking mental events out of their own invisible experience and identifying them as God; they integrate those events into the awareness of a personlike being by using "let's pretend" play; and then they shape their own interior world . . . they learn to react emotionally to that being, as if that being were alive in an ordinary way right now.[16]

In other words, a Vineyard member may survey her internal, spontaneous mental states—images, feelings, intuitions, and so on—and select from among them those to weave into a story about what God is saying to *her*. This is a voluntary selection of internal mental *props* around which to shape a personal narrative.[17] And after this selection, there is a still choice as to whether to "believe" they count as God's communication.

All of this contrasts with factual belief. Recall:

If you factually believe it, you can't help believing it.

As we have seen, this applies to factual beliefs about such things as what your name is, whether there are apples in the fridge, what state Houston is in, and so on. You didn't acquire those factual beliefs by choice, nor could you discard them voluntarily. Furthermore, when you are unsure whether something is the case—say you aren't sure there are apples in the fridge—you can't just *choose* to factually believe that there are: any mental state resulting from such a choice wouldn't be factual belief in the operative sense. As Neil Levy and Eric Mandelbaum put it, "I cannot directly decide to believe that today is Wednesday."[18] Factual beliefs, again, are the cognitive bedrock on the basis of which you choose other things, including other attitudes that can be chosen; they are not under direct voluntary control themselves, as formalized in the definitions Involuntariness 1 and Involuntariness 2 in the last chapter.

We have already reached a conclusion worth noting. There exist religious "beliefs" in the minds of an extensively studied Christian group that are voluntary and hence do not satisfy the first principle of factual belief. We are well on our way to showing the *existence* of a different religious attitude type, one that is voluntary.

*   *   *

In a recent paper titled "Seeking the Supernatural," Michiel van Elk and I distinguish *general* religious credences from *personal* religious credences.[19] This distinction takes attitude type as given and makes a further division in terms of content. General credences have culturally widespread contents, like *God exists* or *God speaks to people*. Personal credences, however, indexically refer to the believer herself (they have indexical constituents like *I, me,* or *my*) that cannot be straightforwardly derived from general credences that the believer already has. So personal credences have contents like *God visited <u>me</u> in the hospital* or *God's voice told <u>me</u> to go on a mission to Mexico*.[20]

This subsection so far has shown that personal credences in the minds of Vineyard members are largely voluntary (e.g., one chooses to "believe" *God spoke to me as I was praying about my career*). It is a further question whether the general religious "beliefs" Vineyard members have are voluntary. On this issue, the evidence is less direct but still compelling. I start with the broader category of American Protestantism and then focus again on the Vineyard.

At a party in Philadelphia many years ago, I once asked a then-recent convert to Protestant Christianity why he adopted his "beliefs." His answer: "I wanted that as part of my life." That suggests he *chose* general Christian "beliefs" for their effect in his life: he could just as well *not* have chosen those "beliefs" had he not wanted those effects. I think his outlook is representative.

In sociological research on conversion, the following pattern emerges. The majority of conversions to a religion (like Christianity) are gradual, deliberate, and done for practical reasons; they aren't blitzes like Paul is said to have experienced on the road to Damascus. Many people are attracted to dramatic conversion *stories*—and often recount their own conversions this way. But in fact most would-be converts observe and weigh the largely social costs and benefits of belonging to a particular religious group. Once the case appears strong that belonging is worth it, the convert accepts the general "beliefs" of the church as well.[21] There may be moments of sudden "revelation" along the way—intense feelings of communion, and so on—but these are mysterious enough to be as open to interpretation as the "cherry-picked" mental events that undergird Vineyard members' personal credences. So voluntary control over general "beliefs" works indirectly: one chooses to belong to a certain religious group, and being in that religious group *socially* constrains one to have and profess certain "beliefs." There might seem to be a lack of choice since many religious groups *require* a profession of "belief," but that's not involuntariness in the senses defined in the last chapter, which are the ones relevant to classifying cognitive attitudes. Insofar as one can choose to leave the group, there is a choice of whether to "believe" that doesn't obtain for factual beliefs.[22]

To return to the Vineyard, there are two indicators of its members' voluntary control over their general religious "beliefs."

First, they *talk* about having a choice—a choice over whether to accept Jesus. Luhrmann writes: "At the church Arnold founded, people . . . talk about deciding that it was time to *choose* Christ."[23] Such talk may seem to presuppose that Christ exists, with the choice being about whether one follows Him. But if one were confident in advance that Christ existed, whether to follow would not be a difficult choice. So it is most plausible that this "choosing Christ" is in the first instance a choice to "believe" that He exists and is God.

Second, both leaders and members of the Vineyard deliberately *shape* the portrait of God they adopt. Many of them regard God as a "best friend," and they hold that "believing" in a God who has the right characteristics will bring joy into their lives.[24] So they *modify* their ideas of God to bring emotional benefits, as Luhrmann recounts:

> They believed that if you could bring yourself to believe genuinely in a loving God, your life would reflect the resilience of someone who believes deeply that he or she is loved. Rachel remarked, "I feel like everyone has a different notion of who God is. All are equally supported by the scriptures. What happens is that you reach a point where you feel like God's not responding or something's not going well in the relationship. Then you realize you think of God as being someone who's angry or unforgiving or whatever. So then *you realize that you have to modify it.*" As a result, evangelicals support an ever more thriving community of Christian therapists who described their primary task as working with someone's inner God-concept.[25]

So it's not only the case that people's general religious credences about what God is like *are* under voluntary control; it's also true—at least in the Vineyard—that many people are metacognitively *aware* that such voluntary control exists and are able to talk about it to therapists and anthropologists. They think of themselves as being able to choose a different God conception, one that (with enough practice) will lead to happiness.

To put this all together, there is the following evidence for the voluntariness of many Vineyard "beliefs." First, Vineyard personal credences—those "beliefs" that involve *I* or *me*—are often acknowledged to be a matter of choice. Second, one's adopting general religious credences is largely a function of which church or religion one chooses, where that choice is influenced by such factors as perceived social benefits. Third, Vineyard members and other evangelicals commonly frame becoming Christian—and hence adopting general Christian credences—as a "choice." And fourth, Vineyard members and other evangelicals deliberately (re)shape their internal descriptions of God to achieve therapeutic/self-help ends.

In all of this, we shouldn't lose sight of *how* religious credences are chosen, which reveals an asymmetric relation between factual beliefs and religious credences—just as with factual belief and imagining. As pointed out, people join a church because of social and other benefits; this is a manifestation of the voluntariness of religious credence. But it is more accurate to say that people choose to join churches because of what they factually believe the benefits will be. A would-be convert observes things about a church (friendly people, good music, short sermons, pleasant coffee and conversation after, many potential romantic partners), and on the basis of the factual beliefs she thus forms about such things, combined with her preferences, she decides to join the church—and in deciding, takes on a host of religious credences by choice. So factual beliefs about a church (and about the consequences of joining) are cognitive inputs into the choice to hold many religious credences. Thus, religious credences, like imaginings, differ from factual beliefs not only in that they are chosen at all but also in that factual beliefs represent a significant portion of the information on the basis of which one chooses them. This asymmetry is crucial to understanding factual beliefs, and by contrast, secondary attitudes. Factual beliefs are unchosen cognitive bases for choosing secondary cognitive attitudes, including religious credences.

2.2. The Compartmentalization of Religious Credence

Recall that factual beliefs help guide behavior whenever one acts in the world.[26] They're not compartmentalized; they are, rather, operative as a class in all settings. Of course, we don't usually notice them because most factual beliefs are obvious enough to escape notice, yet they're still doing work. You go to the fridge because you want an apple. But more accurately: you go to the fridge because you want the apple that you factually believe is in there. You turn left because you want to go to your house. More accurately: you turn left because you want to go to the house that you factually believe you own on Smith Lane. Examples can be multiplied.

They show that one doesn't normally think *about* one's factual beliefs; rather, one thinks *with* them to choose actions that accomplish one's goals. And this is true in *any* situation. Hence: *factual beliefs guide action across the board.* An extension of this point is that people continually *rely* on their factual beliefs (and on things being as they describe): if you factually believe that *p*, you are generally ready to behave confidently as if the state of affairs described by *p* is real.

Can the same be said for religious "beliefs"?

Consider these observations from Luhrmann's more recent book *How God Becomes Real*:

> Devout modern Christians talk constantly about not being faithful enough. They bemoan how hard it is to keep God's love at the front of their minds. They complain about forgetting about God between Sunday services. They apologize for not being able to trust God to solve their problems. I remember a man weeping in front of a church over not having sufficient faith that God would replace the job he had lost. When you pay attention, you can see that church services are about reminding people to take God seriously and to behave in ways that will enable God to have an impact on their lives: to pray, to read the Bible, to be Christ-like. And then people say that they go back home and yell at their kids and feel foolish because they have forgotten that they meant to be like Jesus. They report running out of time to pray. They confess that they do not behave as if God can help them. They worry that they do not really understand or commit as they should.
>
> In fact, when you look carefully, you can see that church is about changing people's mental habits Sunday by Sunday so that they feel that God is more real, more relevant, and more present for them—so that they believe more than they did when they walked in and hold on to those beliefs a little longer after they walk out. It is one of the clearest messages in Christianity: *You may think you believe in God, but really you don't. You don't take God seriously enough. You don't act as if he's there. Mark 9:24: Lord, I believe; help my unbelief.*[27]

I want to focus on two themes from this passage.

The first is that modern Christians (including Vineyard members) confess that they behave as if God doesn't exist *when it's not Sunday or other sacred times.* That pattern is common enough that there's a phrase for it: "once-a-week Christian." Many of them *wish* they would more consistently "take God seriously," but the fact that that's a wish shows that it's contrary to how things usually are (most of the week: *you don't act as if he's there.*).

The second theme is the often unsuccessful *effort* people put into acting on their religious "beliefs." That effort is a sign that those "beliefs" don't work like factual beliefs.[28] A One-Map Theorist looks at religious rituals and says, "See, those people straightforwardly think their religious ideas are true." But that view ignores Luhrmann's point that many rituals are designed (in some sense and among other things) to *strengthen* "belief" so that people "hold on to those beliefs a little longer." Such deliberate strengthening would make little sense if religious "beliefs" were straightforward factual beliefs. Factual beliefs (with contents like *dogs have teeth* or *Julia Roberts acts in movies*) do not need strengthening; we just rely on the world's being as they describe. If they are strengthened in any sense, it is by evidence rather than ritual.[29] So rituals and practices designed to bolster "belief" are evidence that people's religious "beliefs" are *not* factual beliefs. To apply this point to the Vineyard: the fact that it takes people ritualized effort to act as though God exists when it's not Sunday is evidence that the default psychological function of the Vineyard credences (that is, how they tend to work in point of psychological fact) is to guide behavior *in sacred times and places*—while lying mostly dormant otherwise.

An illustration of Vineyard compartmentalization comes from Bialecki's discussion of how members exhibit "demonic attacks."

> One of the interesting things about the sort of demonic attacks that trigger deliverances is they seem to always occur in charismatically intense settings. . . .
>
>   Other than the prophet [mentioned above], I never heard of anyone who claimed to have encountered a full demonic manifestation "in the wild," that is, outside a charismatic service or a collective session of charismatic prayer.[30]

In other words, Vineyard members may act in ways that suggest demonic possession *when they are in a sacred setting* but not typically otherwise. Not only do people fail to behave as though *God* exists outside sacred times and places (as Luhrmann observes); they also don't behave as though demons exist outside religious settings (as Bialecki observes). It is as if the supernatural beings go on holiday when it's not church time or when other church members aren't present. The compartmentalization of credences about demons in particular is significant because the *contents* of credences about demons would suggest that demons are *more* likely to be present "in the wild" since demons are supposed to be weakened by God's presence. Why, if demons are afraid of God, do their manifestations only show up in settings where God is supposed to be maximally present?[31] One could, of course, dream up a story for why this might be so (e.g., the demons want to challenge God, etc.), but Bialecki never mentions a story of this kind. The absence of serious demonic affliction outside "charismatically intense settings" is a surreptitious expression of a factual belief map that *lacks* demons.

Bialecki offers further insights into how Vineyard members deploy their demon ideas. Sometimes members do gloss daily afflictions as being minor demonic harassments (as opposed to full-blown demonic attacks), but in those cases, naturalistic and supernaturalistic "double coding" is typical:

> Not everyone in the Vineyard resorted to the language of demonic attack when explaining mounting frustrations. Furthermore, those who did rely on demonic accounts were fully capable of "code switching," producing demon-free, quotidian secular narratives of the same events. My sense is this is not a case of people having learned to adopt secular language but an indication there are multiple causal models available . . . these accounts are *context dependent*; a person who has received a shock from the electric coffeemaker may reference it as a demonic attack but still be sure the appliance is electrically grounded the next time it is used. And despite their effective copresence, both accounts are complete on their own.[32]

In combination with the last quotation, this one suggests the following picture of how Vineyard members process their ideas about demons. First, those ideas don't influence "quotidian secular narratives": they are part of an independently existing cognitive map. Second, those ideas guide verbal and other behaviors (demonic manifestations, etc.) in religious or "charismatic" practical settings. They are, as Bialecki puts it, "context dependent."

Let's tie this to the framework developed in Chapter 2. We saw how a brick could be imaginatively transformed in a pretender's mind into a sandwich, but this transformed identity is only operative in guiding behavior during an episode of make-believe. Similarly, an electric shock from the coffee maker can be transformed in a Vineyard member's mind into demonic harassment, but that transformed identity is "context dependent" or only operative in episodes of the religious setting. This illustrates the existence of a religious attitude that does not satisfy the definitions of practical setting independence given in the last chapter (just like imagining doesn't satisfy them).

Bialecki's observations about demon-related behaviors also illustrate how factual beliefs stay operative in guiding people's behavior, even in charismatic settings. When one parses the electric shock as demonic harassment, one is still being guided by the factual belief *that I was shocked by the coffee maker*. The electric shock, then, is a *prop* that one uses (voluntarily) to generate credences that encode a supernatural narrative (just as the brick was a prop used in generating a story about sandwiches in the game of make-believe). So people's factual beliefs keep track of the objects in their environments that are to be used as props. Regardless of whether one is parsing the shock from the coffee maker as an electric malfunction or as demonic harassment, one is guided by the factual belief *that I was shocked*. With this factual belief in the background, the religious setting activates religious credences that parse the shock as demonic (and those same credences become deactivated outside the religious setting).

What is the religious practical setting? The details vary from sect to sect and person to person, but—similar to the setting of make-believe— three expectations in a person's mind typically constitute that person's religious practical setting, and these expectations are usually shared by other participants, if any are present. First, people expect that episodes of religious activity will have a limited duration in time. And thus, no matter how fervent one is in a given moment, one is aware that the time for acting religiously will end. Consider this passage from religion scholar Cheryl Townsend Gilkes on the therapeutic role of Christianity in Black American communities. After describing how enthused people can be by the Holy Spirit in the church setting, she notes:

> No matter how severe the pandemonium within the church service, I have never witnessed a church service in which every single person's episode of "getting happy" or "shouting" was not resolved, worked through, or

finished before the singing of the final hymn and the recessional. When participants leave, they usually appear as unruffled as they did when they came into church.[33]

Religious behavior thus comes to an end when the service or other sacred time is over, and this is significant given that—from the perspective of doctrinal contents—God is supposed to be present everywhere and always.[34] So compartmentalization is best seen as a function of *attitude* rather than content. Second, people expect there to be certain *cues* at the starts and stops of episodes of religious activity, such as the opening prayer or the "singing of the final hymn and the recessional." Overt cues are not always employed: it may be enough that a group of people has come together. But the fact that standardized cues exist across a large variety of cultural contexts—the beating of certain drums, the call to prayer, the ringing of particular church bells—makes it worth building the expectation of cues into our conception of the religious practical setting. Third, the religious practical setting includes the expectation that some objects, places, and events will be assigned—in a given episode—identities other than what they are factually believed to be; in other words, there is the expectation that a second, typically supernatural, set of identities will be assigned. Cheap wine becomes Christ's blood (while participants are aware that it is cheap wine); auditory imagery in one's head becomes the voice of God (while one is aware that it needn't be construed as "from God"); and the empty chair at the dinner table (at least among Vineyard practitioners) becomes the place where God is sitting down to eat with you. Importantly, the transformed sacred objects can return quickly to their original, mundane identities *outside* the sacred time and place. Twentieth-century anthropologist E. E. Evans-Pritchard writes of the stones used in a ritual in a preindustrial society: "Some peoples put stones in the forks of trees to delay the setting of the sun; but the stone so used is casually picked up, and has only a mystical significance in, and for the purpose and duration of, the rite."[35] In sum, religious practical settings are constituted by participants' expectations of limitedness in time and space, of characteristic cues that mark the limits, and of alternate assigned identities that obtain within the relevant limits.

Let's relate this to the broader aims of this chapter. Since I'm defending a particular Two-Map Theory that applies to the Vineyard, I am attempting to show that *many* of the religious "beliefs" in the heads of Vineyard members lack the four main features of factual beliefs. In this subsection, I am arguing that Vineyard members have "beliefs" that contrast with factual beliefs in that they are compartmentalized or *practical-setting-dependent,* and the characterization just given outlines the structure of the setting, with its three expectations, to which religious credences are compartmentalized. So let's apply the last paragraph to what went before. When the prayer groups Bialecki discusses meet, there are fairly clear starts and stops to the religious activities (expectation 1); certain cues, like the playing of a particular song on the acoustic guitar, signal the start of the religious activity, while others, like a "closing" prayer, signal the end (expectation 2); and participants also expect that certain objects and events will be assigned different identities from what they have in everyday life, like when an emotional outburst is parsed as a demonic affliction (expectation 3). In the last subsection, we saw that many Vineyard "beliefs" are voluntary; here, we see that many are compartmentalized.

The last chapter also explained how factual beliefs are *the practical ground* of imaginings. That means that when one is in the setting of make-believe play, factual beliefs themselves represent that one is in that setting (one factually believes one is playing make-believe). Factual beliefs are also the practical ground of religious credences: that is, one's factual beliefs track situational features that determine whether one is in a sacred practical setting. Whether it's Sunday, for example, is something you have factual beliefs about (and you can't just decide to believe it's Wednesday!). Whether the building you are in is a church is also something you have factual beliefs about, even or especially if you're in a gym that's been repurposed for the day. Whether the group of people around you are church members is also something you have factual beliefs about. So, in representing *specific practical settings,* factual beliefs in part constitute them, including religious settings.[36] Metaphorically, it is as if factual beliefs say to religious credences: "Okay, we're now in the sacred setting. The cues have been given, so now you get to structure sacred and symbolic bodily movements. But we (factual beliefs) will still keep track of the ordinary features of the physical environment and will let you know when the sacred time is over, at which point you no longer will guide action." Factual beliefs are thus used to *manage* one's secondary cognitive attitudes, including religious credences: they help activate and deactivate them. So this is another way factual beliefs are basic in relation to religious credences: one's factual beliefs about whether it's Sunday (or whatever) activate religious credences, but religious credences play no analogous role in relation to factual beliefs since factual beliefs as a class are always operative in guiding behavior anyway.

There is one final and important way to see that factual beliefs are fundamental and uncompartmentalized. Even when people are engaged in religious activity (in church or in another sacred setting), they rely on their factual beliefs to operate effectively in the physical world. And this point carries over to the Vineyard. Bialecki emphasizes that Vineyard services typically include PowerPoint projections of worship lyrics so that congregants can sing along. So the person operating the slides, who is presumably also worshipping, must know which PowerPoint files correspond to which song, which button dims the projector and which button lights it up, where the spare projector bulb is stored, and so on. And factual beliefs encode the awareness of these basic facts—the sort of factual beliefs that could easily be shared by religious devotee and atheist alike. So, again, factual beliefs operate constantly even in the worship setting to enable action in the physical world. Others worshipping must also track things like how long it's been since they took their insulin shot or took their child to the bathroom. All of these things require the continual operation of factual beliefs about time, medical conditions, and basic needs and habits of one's offspring—factual beliefs one would rely on whether one was worshipping or not. In short, though much ordinary behavior is guided by factual beliefs and not at all by religious credences; behavior in sacred settings is guided by both religious credences *and* factual beliefs.

So, again, factual beliefs as a class are always operative, while religious credences are only sometimes operative (and as a second map layer over the factual beliefs): that is a crucial disparity between the attitude types— one that in part constitutes religious credence as a secondary cognitive attitude.

2.3. Limited Cognitive Governance for Religious Credence

The next phase of the argument that many Vineyard "beliefs" are secondary cognitive attitudes is more technical but just as significant. The point to carry over from Chapter 2 is that factual beliefs have what I call *widespread cognitive governance.* Here, I show that many of the religious "beliefs" (religious credences) in the minds of Vineyard members do not have that, which makes them also secondary cognitive attitudes. What I'm showing here is that, in the two-map cognitive structure that guides Vineyard behavior, one map cognitively governs the other, but not vice versa.

Let's rehash what cognitive governance is. Roughly, saying that a set of mental states *A* cognitively governs a set of mental states *B* means that the representations in *A* are used to draw inferences from elements of *B* to new elements of *B*. Otherwise put, *A* is the informational background that supports inferences among elements in *B*. Recall how factual beliefs cognitively govern imaginings: if Kevin imagines lightning striking a tree, he probably next imagines the tree being burnt (or something similar). What facilitates his transition from imagining the lightning strike to imagining the charred state of the tree? That transition doesn't follow by logic alone; some background information is needed. The answer is that it's his factual belief *that lightning burns wood.* Furthermore, this governance is asymmetric,[37] meaning factual beliefs are not also governed *by* the other cognitive attitudes, like imagining (e.g., if Kevin merely fictionally imagines lightning hitting the tree in front of him, he doesn't just start believing that it's on fire; that's because the governing goes one way and not the other). I summed this up in the principle that *factual beliefs guide inferences in imagination.* But again, the point also extends beyond imagining. That is, the cognitive governance of factual beliefs is *widespread*: factual beliefs govern not only imaginings but also other factual beliefs and secondary cognitive attitudes of every sort, like suppositions, hypotheses, guesses, assumptions for the sake of argument, and so on.[38]

With that as background, this subsection argues that many Vineyard "beliefs" fall in with the other secondary cognitive attitudes in two ways.

(1) These Vineyard "beliefs" do not govern factual beliefs.[39]

(2) These Vineyard "beliefs" are governed *by* factual beliefs.

We can see support for (1) by examining a big-picture quotation from Bialecki. Here's some background to the quotation: Bialecki's term "diagram" refers to a cluster of practices, patterns of thought, ways of identifying phenomena, and expectations that are characteristic of a certain social setting for a certain social group.[40] Thus, *A Diagram for Fire,* as an ethnography, illuminates the "Vineyard diagram"—the diagram at work in the Vineyard community. As indicated, Bialecki discusses *prophecy* as a component of this diagram. Interestingly, at one place, he mentions how one pastor allowed church members to think of their internal premonitions *either* as prophecy (a supernatural frame) *or* as intuition (a psychological and hence naturalistic frame), which further exemplifies the double coding often discussed. Here's the interesting part for present purposes. Of this pastor's allowance of double coding of intuition/prophecy, Bialecki writes:

> This tendency to shift to constituent framings and let the [Vineyard/religious] diagram collapse is at times linked to how Vineyard believers must live in a secular world infused with countless other religious possibilities, *including the possibility of there being no religion and no transcendence at all. . . . Such awareness, however, is not just an abstract cognitive frame, a simulation or model of the beliefs of other people; it is an embodied and unconscious sense of how one can maneuver in the consensual world created by this cohabited plurality.* It is not surprising at all that when a more openly charismatic diagram decoheres, the next stable state that it collapses into should be a set of immanent relations in which the miraculous and God are not immediate forces.[41]

So even the most charismatically oriented Vineyard members, like this pastor, can go about the world in a way that does *not* incorporate the charismatic or supernatural ideas for which the Vineyard is known. Moreover, they are not living and acting in these ways merely to humor non-Vineyard members. Rather, the "charismatic diagram decoheres" and "the miraculous and God are not immediate forces." This shows that Vineyard members have a basic class of representations of the world that is unpenetrated by their charismatic and supernaturalistic "beliefs," and this class (map) allows them to act in an entirely naturalistic, nonreligious way much of the time. For this to be true, the specifically religious Vineyard "beliefs" must *not* cognitively govern that more basic class, since otherwise, the more basic class would be infused with the supernatural, contrary to what Bialecki reports. Vineyard members can return to the Vineyard diagram when the time is right, but there is still a separate "stable state" that involves interacting only with mundane, quotidian entities and events.

This is similar to the perspective Bialecki offers on Vineyard attitudes toward demonic possession, quoted above. Recall that Bialecki maintains that Vineyard members who offer demonic accounts of their troubles can also give naturalistic accounts, like that the coffee maker that shocked them wasn't grounded. And they rely on those naturalistic accounts in non-Vineyard contexts (practical settings). Furthermore, both demonic and naturalistic accounts are "complete on their own." That claim of completeness has some complications. But let's focus on one side of it, which is straightforward. To say that the naturalistic account of certain events is complete in itself implies that, in my terms, it rests on factual beliefs that are *not* governed by religious "beliefs" about demons. So when Vineyard members (who might at other times talk of demons) are in the naturalistic "stable state," they do *not* rely on religious credences about demons to make inferences, say, about whether the coffee maker is grounded, whether there is a loose wire, whether it is touching something else that is electrically malfunctioning, and so on. Other factual beliefs about electricity, metal conductance, or one's observations undergird those inferences since the class of factual beliefs cognitively governs itself. The factual belief explanation is, again, "complete" on its own—ungoverned by religious credences about demons.

Now let's turn to another cluster of credences: religious credences about petitionary prayer. Three interesting facts about petitionary prayer are covered in both Luhrmann's and Bialecki's ethnographies. Together, they point to a lack of cognitive governance on the part of religious credences.

The first is that Vineyard members engage in petitionary prayer *a lot*. They pray for relief from their medical ills and for help with a range of other issues, from success on a university exam to success in a job search. A One-Map Theorist would look at this first fact and say, "See, they simply and straightforwardly think that praying to God will help solve their problems." But the second and third facts show why that perspective is simplistic.

The second fact concerns what people *don't* pray for. Generally, though dramatic exceptions exist, people avoid praying for events that are impossible from a naturalistic standpoint. People may pray for God to help them find a competent, reasonably priced mechanic who can fix their car.

But they don't pray to God to make a working carburetor appear ex nihilo. Why not? The explanation, I submit, is that people's factual beliefs about how working carburetors can get there are ungoverned by religious credences—no matter how powerful those credences portray God as being. Bialecki tells a story of how one young woman was suffering from medical problems and ensuing debt, but despite praying for her back pain to go away and for job opportunities to open up, "no prayer . . . was offered about her debt."[42] In other words, no one asked God to wipe away debt without the occurrence of changes in the natural world that would be intelligible to a nonbeliever. No one, for example, asked God for a direct deposit. The prayer group ended up pitching in their own money and *subsequently* attributing that action to God. That is, they made a natural change in the natural world of the sort that any group of friends might make to help someone with financial difficulties, and then they formed the religious credence that *that* had been God's handiwork. So Vineyard members don't typically pray to God for things that would obviously violate what their factual beliefs entail is possible. Rather, they usually pray for God to make things break their way when things are uncertain, where uncertainty is constituted by what factual beliefs leave undecided (e.g., whether the woman will find a job soon). For that to be the case, those factual beliefs, which *constrain* the possibility space, must not be subject to inferential alterations that integrate the contents of religious credences about divine interventions (otherwise much *more* would seem to be within the realm of reasonable possibility). All this supports point (1) above: Vineyard members' factual beliefs are not cognitively governed by their Vineyard "beliefs."[43]

This second fact, furthermore, coheres with Justin Barrett's experimental research on petitionary prayer. The title of Barrett's paper is already suggestive: "How Ordinary Cognition Informs Petitionary Prayer."[44] His four studies show that Protestant Christians are more likely to pray for divine interventions that would bring about psychological or biological effects than for divine interventions that would involve observable mechanical changes. For example, they are much more likely to pray for God to change someone's mind about where to steer a boat than they are to pray for God to plug a hole in a boat that's leaking. In Barrett's Study 4, furthermore, only 2 out of 70 participants said they would pray for a divine intervention that would violate physical regularities. By way of contrast, 57 and 49 out of the 70 said they would pray for psychological and biological changes, respectively. And those kinds of changes, importantly, are those that are left as *open possibilities* by peoples' factual beliefs. In other words, *people pray for the sorts of things that might happen anyway.* This coheres with my interpretation of the Vineyard evidence: people in Barrett's studies mostly avoid praying for outcomes that would directly challenge or cause revision to their factual beliefs about how the natural world works. This pattern is not a hard-and-fast rule, since, in times of desperation, people will pray for almost anything.[45] But, as Georges Rey points out, even very religious people are more likely to pray for a cancer to go into remission than for an amputated limb to grow back.[46]

The third fact is that, when Vineyard members pray to God for help with a problem, they *also* do the nonreligious activities that (from a factual belief standpoint) would solve that problem. They may pray to God for relief from a medical problem, but they will still see the doctor, get second opinions, take prescribed medicines, and do the usual requisite nonreligious activities. Or they may pray for success in a job search but still submit resumes and fill out job applications. In other words, despite what they might sometimes say, they do not *rely* on its being true that God will help them.

As Luhrmann writes, "They said that they felt embarrassed to feel that they should study in addition to praying they would pass their exam—but they still studied."[47] Concerning prayer for healing, Bialecki writes:

> This providential form of healing is exemplified by a common type of healing prayer, in which the petitioner may bring about the success of a medical procedure by, for example, requesting that God guide the hands of the surgeon; similarly, someone may pray that naturally existing bodily capacities for healing be divinely catalyzed for a quick recovery.[48]

In other words, Vineyard members don't expect God's help to *replace* the things they have to do to solve their problems; rather, the prayer to God is for a bit of extra push in what one is doing and what one thinks is happening anyway. To relate this to the present topic, this fact shows that Vineyard members have a stable picture of how things generally work in the natural world that guides their practical actions. Luhrmann calls this their "everyday frame" (as opposed to their "faith frame"), and I call it factual belief. God's intervention ends up being credited post hoc as a gloss on any favorable naturalistic outcomes. But one still works toward desired outcomes in ways that show consistent reliance on the everyday frame.

To return to the main thread, we now have substantial evidence for (1) (that religious credences do not cognitively govern factual beliefs). The completeness and separability of nonreligious explanations, the ability to double code, the limitations on what people pray for, and the nonreliance on God's help (despite having prayed for it)—all of these reveal the existence of representations about the world and how it works that are unaltered and hence ungoverned by religious credences. Despite their having plentiful religious "beliefs" about prophecy, demons, and petitionary prayer, the factual beliefs of Vineyard devotees (like other Protestants) are generally unaltered by those religious "beliefs." One's factual beliefs supply events and problems for religious credences to be *about,* but the layer of credences adorns, without inferentially altering, the fabric of factual beliefs.

We now come to the complication alluded to above, which is relevant to point (2) (that religious credences are governed *by* factual beliefs). The complication prima facie tells against (2), but more detailed consideration of the relevant evidence supports it. When discussing naturalistic versus demon-invoking explanations of a shock from an electric coffee maker, Bialecki said that both accounts were "complete on their own." I took this as supporting evidence for point (1); namely, that religious credences do *not* govern factual beliefs: since the factual belief map is complete on its own, it is not governed by religious credences.

One might also read Bialecki as implying that factual beliefs don't govern religious credences *either*: both maps are "complete" in themselves, so neither governs the other, which would go against point (2). I can't say whether Bialecki intends that conclusion, but if he does, that particular stance would be undermined by some basic considerations he might have overlooked. (And such overlooking would be no surprise since the operation of factual beliefs is so humdrum that most people overlook it.) Say, to extend his example, a Vineyard member was shocked by a coffee maker and formed a religious credence *that a demon caused the coffee maker to shock me.* If another Vineyard member asked the victim when the shock occurred, she would, presumably, think about what day it had happened and then report an answer. If the shock happened on a Tuesday, she would say the demon harassed her on Tuesday. But drawing that inference relies on a factual belief about *when* the shock happened. So the inferential structure looks like this:

RELIGIOUS CREDENCE: *a demon caused the coffee maker to shock me*
FACTUAL BELIEF: *the electric shock happened on Tuesday*
RELIGIOUS CREDENCE: *the harassment from a demon happened on Tuesday*

This inferential sequence shows the typical cognitive governance of factual belief. The information that allows the inference from the first religious credence to the second is encoded in a factual belief that would *also* operate in the everyday frame.

Factual beliefs also govern religious credences in supplying *topics* for people to have religious credences *about*. For example:

FACTUAL BELIEF: *I was shocked by the coffee maker*

Then, in a religious frame of mind, one might invoke general religious credences to infer further religious credences about the shock:

FACTUAL BELIEF: *I was shocked by the coffee maker*
RELIGIOUS CREDENCE: *demons cause daily tribulations like events that cause pain*
RELIGIOUS CREDENCE: *a demon caused the coffee maker to shock me*

This sequence would be the run-up to the earlier one. Again, we see that basic information encoded in factual beliefs supports inferences among religious credences. So, although I am usually in agreement with Bialecki's analyses, I have to say that his suggestion that religious explanations are "complete" in themselves is misleading since, if they pertain to everyday events *at all,* like electric shocks from a coffee maker, they will draw on information encoded in factual beliefs.

Examples like the (demonic) coffee maker shock can be multiplied to establish (2) that factual beliefs govern religious credences in the sense of governance defined in Chapter 2. If one does poorly on a test, it may

be because demons interfered *or* because God didn't want that person going into that academic field. But either way, the factual belief about one's
performance (e.g., with content *that I got a C*) informs inferences drawn
among the religious credences. And so on for other examples.

Let's visualize it this way: factual beliefs are one's basic map of how
things are in the world, and one's religious credences are a transparency
that lies on top of it. The underlying factual belief layer is not redrawn by
the transparency, though the images on the transparency may obscure
some of its details.[49] But the images and information from the factual
belief map layer can mostly be seen, even when the religious credence
map layer is on top. Furthermore, information in the factual belief map
layer is used to fill out what appears in the religious credence map layer so
that the supernatural entities and events depicted in the religious credence
map layer appear to be *about* entities and events in the factual belief map
layer. In other words, the factual belief map helps govern what gets drawn
on the transparent layer. Then, when one exits a religious practical setting,
one typically stores the religious credence transparency away and lets one's
actions be guided by factual beliefs, as we just saw.

2.4. Vineyard "Belief" and Evidential Vulnerability

Chapter 2 showed that factual beliefs are vulnerable to evidence: they
tend to get extinguished (involuntarily) when the person who has them
cognizes contrary evidence. You might have thought, for example, that
there were almonds in the cupboard, but when you open the door and see
the cupboard empty—*poof!*—that factual belief is extinguished. There
are, of course, many imperfections in the relation between factual beliefs
and evidence due to biases (motivated or unmotivated), lack of awareness
that something is evidence, thinking something is evidence when it isn't,
forgetting, and so on. But in general, people rationally update their basic
internal models of the world in light of incoming evidence. And this feature of factual beliefs helps differentiate them from secondary cognitive
attitudes, which, as we saw, gives us our last principle: *factual beliefs respond to evidence.*

Do Vineyard "beliefs" behave like factual beliefs in this respect? Here, I
give four ethnography-based arguments that a great many of them do not.

The first is that Vineyard members—along with other Christians—often *say* they hold their "faith" either without evidence or even in the face of contrary evidence. (Faith, as I understand it, is a spiritual orientation in life that typically *includes* descriptive "beliefs," such as "belief" that God exists, that He is loving, and so on. So, if one holds faith without or contrary to evidence, one holds those "beliefs" without or contrary to evidence as well.[50]) Luhrmann, for example, reports speaking with a Vineyard member who had been struggling with a mental illness and who often prayed for help but rarely saw improvements in her condition. How, Luhrmann wondered, could this person "believe" that God was there for her in the face of such counterevidence? This is the answer:

> Sarah had learned to value hope more than outcome. Early in our conversations she had quoted Hebrews to me: "Faith is the substance of things hoped for and the evidence of things unseen." *She said that God becomes more real to her in the absence of evidence at all, because then God is her hope, and her hope feels alive and resilient—whatever happens in her world.* "It's contradictory, but it's exactly right," she told me. "I feel really close when it seems as if he's far away. I think that's the only way you can trust."[51]

In short, not only can Sarah "believe" contrary to (or at least in absence of) evidence; she also sees doing so *as a good thing.* Being out of step with evidence makes the "belief" more hope-like and hence psychologically powerful. That idea, furthermore, is not unique to Sarah or even to the Vineyard. The verse she quotes, Hebrews 11:1, is one of the most famous in the New Testament. And one theme common among many and perhaps most of its interpretations is that "faith" or "belief" (*pistis,* in Koine Greek, which can be translated either way) is something one holds regardless of whether evidence supports it. Other biblical passages, cited commonly and approvingly in Christian church services, portray faith as something that one holds in the face of contrary evidence: Noah has faith that building an ark in the middle of a massive drought will somehow serve a purpose; Abraham has faith that sacrificing Isaac will not leave him without an heir; Doubting Thomas *should* have had faith despite not having seen evidence that Jesus had risen; and so on. So not only *can* "faith" and concomitant "belief"—unlike factual belief—survive in the face of contrary evidence; it also should, and Vineyard members and other Christians say as much.[52]

The second argument appeals back to the voluntariness of Vineyard credences. Recall that Vineyard members say they have a "choice" about what to "believe." This latitude to choose one's religious "beliefs" does not *logically* contradict the idea that they are evidentially vulnerable, but it sits ill with it. Generally, the higher degree of evidential vulnerability a mental state type has, the more it will be constrained by (what one takes to be) evidence for or against its contents. So if an attitude has a high degree of evidential vulnerability, as factual belief does, it will not have the latitude to be under voluntary control. Conversely, an attitude like imagining is not evidentially vulnerable, and it is to a large extent under voluntary control. So, in the space of cognitive attitudes, there is a continuous trade-off between (i) being more evidentially vulnerable and less voluntary and (ii) being more voluntary and less evidentially vulnerable. Factual beliefs are highly evidentially vulnerable and not voluntary at all; for contrast, one has some voluntary control over one's acceptances in a context, but these acceptances are fairly evidentially constrained; hypotheses are similar, but they perhaps allow for more voluntary control; suppositions, it seems, are even less constrained by evidence and hence more open to voluntary control; and so on, until we get to whimsical fantasies, which are completely unconstrained by reality-based evidence and very much open to choice. Thus, to the extent that Vineyard "beliefs" are voluntary, as we have seen they are, they will also be in the less evidentially vulnerable region of cognitive attitude space.

The third argument appeals to the fact that Vineyard members often admit *doubt* about their religious "beliefs." Not only are they aware of the doubts of nonbelievers; they also admit—in the right circumstances—their *own* doubts about the invisible supernatural being that they worship. Continuing a trope encountered earlier, one of Luhrmann's informants expressed his doubts about whether his apparent spiritual experiences were genuine by saying, "Maybe it's a spiritual experience; maybe it's a lot of caffeine."[53] And in the last chapter of her ethnography, Luhrmann explains that Vineyard members are generally not in a state of psychological certainty: "The God described in these pages, the vividly human, deeply supernatural God imagined by millions of Americans, takes shape out of an exquisite awareness of doubt."[54] Toward the end of the same chapter, Luhrmann invokes Coleridge to describe how Vineyard "belief" and practice require "suspension of disbelief," which I take to be a bracketing of *doubt* about what they "believe." And her observations on these fronts cohere with Joshua Brahinsky's claim that modern American evangelicals experience "ontological anxiety," meaning they are plagued by doubts about the reality of the entities to which their religious "beliefs" commit them.[55] In short, Vineyard members and other American evangelicals commonly "believe" that p while simultaneously doubting that p—where p is a core religious commitment.

This is a different kind of coexistence from the one discussed in the previous subsection. There, we saw that Vineyard members had coexisting explanatory strategies concerning the same phenomenon: for example, one according to which an electric shock was just an accident and another according to which it was demonic harassment. In this case, however, Vineyard "beliefs" coexist with doubt about the contents of those very beliefs. There are two points to extract from this. The first is that having such "beliefs" differs phenomenologically from having factual beliefs. Recall Dan Sperber's line about factual belief: from the point of view of the believing subject, "factual beliefs are just plain 'knowledge.'"[56] In other words, if you factually believe that p, then it *seems to you* like you know that p (even if you happen to be mistaken). For example, if you factually believe your bedroom has windows, then it seems to you like you know your bedroom has windows (and you're probably right). You might entertain skeptical scenarios in an epistemology class. But such exercises are brief and don't constitute the kind of ongoing doubt with which people of faith have to wrestle. Ordinary factual beliefs make a matter seem settled and thus don't usually coexist with deep doubts on that matter. The second point is that such doubts usually encode awareness that evidence is poor, lacking, or even contrary when it comes to one's "beliefs." Luhrmann spends much of her penultimate "Darkness" chapter describing how Vineyard members wrestle with God's absence, unanswered prayers, and life calamities that thwart what they earlier professed God wanted for them. They have various strategies for calming the ensuing doubt, such as saying that unanswered prayers are God's way of teaching them something. But whether or not such strategies are effective, the present point is that Vineyard "beliefs" are, if needed, well able to coexist with doubt and hence with the kind of poor evidential situation such doubt encodes. More abstractly: doubt that p occurs when one's mind has cognized evidence contrary to p, so if "belief" that p coexists with extensive doubt about p, it is likely that that "belief" is not evidentially vulnerable. The result of such ever-present doubt is that much Vineyard speech about "belief" seems paradoxical on the first read. Luhrmann quotes one Vineyard member as saying of her faith, "I don't believe it, but I'm sticking with it. That's my definition of faith."[57] This would be hard to make sense of on a One-Map Theory, but my framework can explain it: this informant is acknowledging—with stunning metacognitive awareness—that evidence makes her unable to factually believe her articles of faith, but she chooses to have religious credence in them nevertheless.[58]

The fourth argument is that when people do cast aside Vineyard "beliefs," it is not typically because they've discovered that evidence doesn't support them. Rather, they found compelling *social* reasons for leaving the church. Generally, one first leaves the community for social or moral reasons, and *as a result of* deciding to leave the community, one discards (many of) one's Vineyard-specific "beliefs." Bialecki writes:

> Overweening authority is a common reason informants give for leaving similar charismatic groups, though it is rare for this to be a Vineyard problem. For the Vineyard, there are other reasons, such as a certain kind of political friction and social exhaustion. Some of those leaving mention hypocrisy, a sense of people not fully living up to implicit ethical claims regarding their behavior or of being in effect "no different" from the secular world. A different commonly given reason for leaving is resentment against the perceived prohibition of wider contact with the world. . . . This is a perceived and not an actual distance, in which church members feel alienated from the larger world rather than rescued from it.[59]

Social friction, ethical hypocrisy, and a feeling of alienation are social and moral reasons for departing a community. They aren't in any clear way evidence either for or against Vineyard "beliefs," such as those about hearing the voice of God or about the reality of prophecy, which can persist anyway in the face of strong doubt. So the main lever for dislodging Vineyard "beliefs" is not evidence but rather the sense that the social commitment bears too many costs.

It's true that one subset of departing Vineyard members attempts evidence-based arguments against Vineyard "beliefs." But such arguments typically go along with expressions of social dissatisfaction, which appears to be the deeper cause of departure. Bialecki continues:

> Young white men usually begin with a claim that the idea of God "doesn't make sense." Rather than relying on ontological or epistemological grounds regarding the concept of the deity, however, the claim turns on the fact that the nature of the world itself—usually the amount of suffering that is presented as being hardwired into the world through forms of social injustice—is incompatible with the positive, affirming picture of God usually championed by churches like the Vineyards. Quickly, however, the conversation can take another turn . . . behaviors by constituent church members, and often very specific constituent church members, are given as evidence of the dysfunctional nature of the entirety of the church . . . interactions with fellow Christians are the impetus that often drives people away from churches like the Vineyard into nonbelief.[60]

It is hard to determine in such cases how much of the departure is due to (a) the ill fit between evidence and "belief" versus (b) the social reasons. Importantly, *the newly minted nonbelievers may themselves not always know.* But in my view, the best overall theory will attribute causation to (b) more than to (a). As we just saw, many Vineyard members see "believing" despite ill fit with evidence as a *good* thing (I discuss why that might be in Chapter 6). So, in *general,* (b) is more of a factor than (a), which means that a more parsimonious theory of the special case will lean more heavily on (b) as well. So when the "young white men" gesture at the Problem of Evil as an evidential argument and *also* mention their social dissatisfaction, the evidence-based argument is likely a way of giving themselves *permission* for what they wanted to do anyway: leave the church.[61]

To recap these arguments: (i) Vineyard members often say they lack evidence for many of their "beliefs" and even acknowledge contrary evidence; (ii) they hold such "beliefs" voluntarily; (iii) they hold them while simultaneously doubting them; and (iv) when they do discard their Vineyard "beliefs," cognition of evidence appears to have little to do with the process. As converging evidence, this all supports the claim that there *exists* a "belief" type that is not evidentially vulnerable, which means there is a "belief" type that is not like factual belief when it comes to evidential vulnerability.[62]

3. OVERALL PICTURE: RELIGIOUS CREDENCE AS A SECONDARY COGNITIVE ATTITUDE

I made the point in the last chapter that philosophers often characterize belief as "the map by which we steer the ship," meaning beliefs are cognitive inputs into action choice. While this is a useful phrase, it leads to an oversight. Often, other cognitive maps are also used as inputs into action choice: imaginings, assumptions for the sake of argument, acceptances in a context, suppositions, hypotheses—all the secondary cognitive attitudes. This led us to theorize what's distinctive about beliefs in the sense in which philosophers usually use that word; namely, our basic internal model of the world that guides goal-oriented action.

What is distinctive about beliefs in *that* sense—which I dubbed *factual* belief for clarity—is that they are involuntary, are practical-setting-independent, have widespread cognitive governance, and are vulnerable to evidence. Secondary cognitive attitudes, which each have their own idiosyncratic practical settings and uses, lack those four properties.

I have shown here that many Vineyard "beliefs" in the minds of real people also lack those four properties: these "beliefs" are chosen, compartmentalized, lacking in cognitive governance, and not evidentially vulnerable; hence, they are secondary cognitive attitudes. The term I give this kind of secondary cognitive attitude is *religious credence.*

It is clear from the ethnographic data considered thus far that religious credence exists and is widespread. It is alive in the minds of members of the Vineyard church, who in rather striking ways—most notably their penchant for playing make-believe—resemble the hypothetical members of The Playground that we saw in the Prologue. So, to answer the question from the start of this chapter (*What attitude . . . ?*), I say this: for any religious doctrine or story, it is likely that humans at large hold a range of attitudes toward it, since content and attitude vary independently, but one cognitive attitude that is both widespread and strikingly similar to make-believe imagining is religious credence, which is far different from factual belief.

Some questions now arise. *What purpose* is served by having a secondary cognitive attitude of religious credence? What are the *positive properties* of religious credences that differentiate them from other secondary cognitive attitudes like imaginings or supposition? I turn to these questions in

Chapters 6 and 7, which explore the roles of religious credence in defining group identity and activating sacred values. Chapter 5, as promised, will discuss the extent to which people themselves are *aware* of differences in types of "belief" that they and others hold. In the next chapter, however, I want to address another important question. Now that we've seen that religious credence occurs in one contemporary American religious sect, what evidence is there that it occurs in other religions, cultures, times, and places around the world? Is there reason to think that religious credence, as an attitude type, occurs crossculturally and in the past?

CHAPTER FOUR

Evidence around the World

1. IS THE TWO-MAP STRUCTURE JUST A WEIRD THING?

The idea of religious credence implies that much of people's processing of religious ideas is like their processing of the ideas that structure make-believe play. For many readers, this view will be surprising. We might find it unlikely that something so serious as religion, for which people are sometimes willing to die, could be psychologically processed, in basic respects, like the ideas behind make-believe play. If one put this thought forth as an objection, my response would be simple: the objection underestimates how seriously humans take games of make-believe. People devote their lives to pretending—from movies to stage acting to video games—and they make sacrifices in time, effort, and money for these activities. As Johan Huizinga details in *Homo Ludens,* many of humanity's gravest cultural forms, from ritual aggression to courts of justice, have their origins in gamelike play traditions.[1] The surprising quality of my view is thus inherited from the astonishing human characteristic of being able to be deadly serious about play.

Yet my view is, from a different angle, intuitive. The deities people worship are *invisible.* Props—statues, paintings, other elements of ritual enactment, or just empty spaces before which one kneels—are used in their places. This resembles the way dolls, toy props, and reimagined spaces are used in the pretend play of children. This similarity was the point of The Playground parable: dolls were transformed into deity statues with substantial continuity in their use. And when people sacrifice animals to their hungry gods, *they* (the humans) eat the meat, much like a child eats the cookie she pretended to feed her doll.

Chapter 3 established the *existence* of religious credence as distinct from factual belief. Many "beliefs" in the minds of Vineyard members are secondary cognitive attitudes. And importantly, the Vineyard is a large, international, and representative evangelical sect.

That is no small thing. But is the religious credence concept applicable around the world? One line of doubt goes like this. *Sure, in secular, scientific societies with religious diversity, people might have religious credences rather than factual beliefs about deities and other supernatural beings. But that's because those people are exposed to differing viewpoints that sow doubt. In more religiously homogeneous societies, far fewer occasions for doubt about the supernatural arise. So people in those societies factually believe that their gods and other religious supernatural entities exist.* I take this suggestion seriously. After all, Joseph Henrich, Stephen Heine, and Ara Norenzayan argue persuasively that WEIRD people (Western Educated Industrial Rich Democratic)—a category to which most Vineyard members belong—are outliers among humanity at large, even in terms of characteristics that other theorists have considered psychologically basic: susceptibility to perceptual illusions, reasoning (analytic versus holistic), attention (figure versus ground), habits in economic cooperation, and degree of autonomy and individualism (versus collective and kinship-based conceptions of self).[2] Maybe the attitude of religious credence as I have described it is a WEIRD-only phenomenon as well; call this the WEIRD-only hypothesis.[3]

Some anthropologists make assertions that seem to imply this hypothesis. *Their* informants simply take the existence of supernatural agents as real in the same way they take rocks and trees as real—as if the humans they study lacked the cognitive flexibility to distance themselves from their own religious ideas. In his 1937 book on the Azande, E. E. Evans-Pritchard writes the following about their attitudes toward oracles, witchcraft, and witch-doctorhood:

> In this web of belief every strand depends on every other strand, and a Zande cannot get out of its meshes because it is the only world he knows.

> The web is not an external structure in which he is enclosed. It is the texture of his thought and he cannot think his thought is wrong.[4]

This passage seems to confuse two things: (i) the interrelatedness of a set of ideas and (ii) having or not having cognitive flexibility in relation to those ideas. It suggests that the interrelatedness of a set of ideas implies that those who possess them are unable to regard them in any other way than as what must be the case. That implication is false: if it were true, Evans-Pritchard *either* would have been stuck regarding witchcraft as fact (since he possessed the "web" of related ideas), *or* he would not have possessed the very ideas he discusses (in which case, why was he writing about them?). From what I can tell, Evans-Pritchard started to distance himself from such implications later in his career, as we'll see below, but other anthropologists have been inclined to attribute an inflexible mindset to the peoples they study. Robin Horton, notably, claims that in traditional African cultures, "there is no developed awareness of alternatives" to their religious and magical ideas.[5] More recently, Christina Toren writes: "If I am to correctly represent my Fijian informants . . . I should say that they know that their ancestors inhabit the places that were theirs."[6] It is worth adding that philosopher Charles Taylor has argued for much the same about preindustrial Christendom: people back then *could not* conceive the world as being other than how their theistic "beliefs" describe, and the fact that we are able to do so now is what characterizes our current, distinctive "secular age."[7] There is thus a prominent strain in anthropological and historical scholarship that implies that any nonmodern or non-WEIRD humans simply factually believe their supernatural ideas, which in turn would imply that the construct of religious credence I have developed is otiose in relation to those humans.

As I've said, I take this WEIRD-only hypothesis seriously, but that is because it is tempting enough that it is worth the effort of refuting it. I have no wish to dispute the *particular* One-Map claims made by the anthropologists just quoted—some of their societies of interest may be as described, though we should still evaluate the evidence critically. But I oppose the *generalizing* idea that all non-WEIRD humans lack two-map cognitive structures in which religious credences are the secondary layer. On the contrary, two-map cognitive structures, similar to what we saw in the Vineyard, are common around the world and also in a range of preindustrial societies. That's what I argue here: religious "beliefs" that are voluntary, practical-setting-dependent (compartmentalized), lacking in cognitive governance, and evidentially unresponsive occur among people in *many* cultures, Western and non-Western, Christian and non-Christian, industrial and preindustrial, as the range of evidence I discuss here reveals.

As evidence converges, the most parsimonious conclusion is this: *the cognitive flexibility to instantiate distinct cognitive attitudes toward religious ideas is widespread and not just the product of WEIRD cultural contexts or education.* Humans generally are capable of more psychological relations to their supernatural ideas than just factually believing them. And though I grant much variation—in fact, one of the points of my theory is to enable us to describe the variation—religious credence is nevertheless a cultural attractor that crops up all around the world and throughout history.

2. TWO-MAP STRUCTURES AROUND THE WORLD

Recall Vineyard *double coding*, where one person alternately gives the same phenomenon a naturalistic description or (in sacred settings) a supernatural description. The coffee maker's shock is due to a lack of grounding—*and/or* demonic harassment (where this later description comes out around other Vineyard members and rarely otherwise). Importantly, Vineyard double coding is a special case of a worldwide psychological phenomenon known as *explanatory coexistence.*

Explanatory coexistence occurs when one person has two or more distinct—and often incompatible—frameworks for explaining the same phenomenon. Typically, people alternate between the different explanatory frameworks, depending on the context.

In one line of studies, Cristine Legare and Susan Gelman examined how Sesotho speakers in Southern Africa maintained both *biological explanations* for how people get AIDS and *bewitchment-based explanations.* Their participants appealed to the virus to explain AIDS, but they also appealed to the work of malevolent witches. But the study participants did *not* get confused between the two kinds of explanation, as a One-Map Theory would predict. "Rather," Legare and Gelman write, "bewitchment and biological explanations co-exist within individuals."[8]

Legare and other colleagues surveyed multiple studies on explanatory coexistence in cultures as diverse as Mexico and China. They found that coexisting naturalistic and supernatural explanations in three topic areas (origin of species, illness, and death) are common and arise already in children.[9] And Justin Busch and his colleagues observed similar patterns of explanatory coexistence concerning those topics on the Melanesian island of Vanuatu.[10] Tying this research together in another survey article, Legare and Andrew Shtulman bluntly state that "individuals who hold only one explanation of a given phenomenon are few and far between."[11]

In addition to not confusing different explanatory frameworks, study participants in this research typically respond in one of three ways when pressed by researchers to account for how their different explanations fit together. Rachel Watson-Jones, Justin Busch, and Legare call these *integrated, synthetic,* and *target-dependent.*[12] Only with integration do participants try to give unified explanations that incorporate both frameworks. With synthesis (this is a term of art), participants produce two explanations side by side without combining them. And with target-dependent thinking, they switch back and forth, depending on what exactly they're being asked about. This shows that the cognitive maps that represent the different explanatory frameworks are functionally distinct in people's minds since deploying them separately (rather than combining them) is more common.

That explanatory coexistence occurs in diverse cultures and appears among young children undermines the idea that the cognitive flexibility it requires is a rarefied, WEIRD phenomenon. The explanatory coexistence literature strongly suggests that the *capacity* for two-map cognitive structures is species-wide. I don't find that surprising since pretend play is also species-wide, and, as discussed in the Prologue, pretend play implicates two-map structures (along with nonconfusion and continual reality tracking).

So we should not be surprised to find crosscultural evidence of two-map cognitive structures that distinctly process naturalistic and supernatural ideas. Now let's discuss evidence that pertains to the typical characteristics of the layer, when we do find such structures, that houses the supernatural/religious ideas.

2.1. On the Voluntariness of Religious "Belief" Crossculturally

The voluntariness of religious credences is part of what makes three wide-spread and quite familiar religious phenomena possible: *creativity, syncretism,* and *conversion in response to incentives.* Here, I discuss plausible examples of each, though many others could be noted. What's notable is that voluntariness plays a role in explaining all three phenomena.

Creativity. The army of terra cotta warrior statues, rediscovered in 1974 after being buried for over two thousand years, stands about an hour outside the Chinese city of Xian. It comprises over six thousand life-size figures of men and horses who protect the grave of the first emperor of China, Qin Shi Huang. The gravesite required around seven hundred thousand slaves to complete it. The army faces east, the direction from which most invading armies would have entered at the time of Qin Shi Huang's death in 210 BCE. So we can infer that the emperor, in some sense, "believed" that the statues—or perhaps the ghost warriors that would arise from them—would protect his spirit. Why else would he have invested the work of seven hundred thousand slaves on the project, only to bury it out of sight? The present point, in any case, is that such "beliefs" were *creative*: they were without a straightforward precedent, and no evidence compelled them. As far as we know, Qin Shi Huang was the first Chinese ruler to bury statues in this fashion (earlier, the Zhou rulers buried actual people for similar grave-protecting purposes). No evidence or prior tradition *required* him to "believe" the statues would be effective guardians, but if he had the latitude to creatively form that "belief," he also had the latitude not to. This suggests that this creative "belief" formation was voluntary in a way that factual belief formation is not.[13]

Another example of creativity comes from how ideas of gods were sometimes generated in traditional Igbo society in (pre)colonial Nigeria. In his novel *Arrow of God,* which portrays Igbo religion in detail, Chinua Achebe indicates that when a new alliance between villages would come about, the village leaders would "make a god" to have a deity they could worship in unity.[14] Characters throughout the novel use that very phrase and similar ones. For example: "Then one day the men of Okperi *made a powerful deity* and placed their market in its care." "Every boy in Umuaro knows that *Ulu was made by our fathers* long ago." "That was why our ancestors . . . *made the great medicine* which they called [the god] Ulu."[15] What's striking is that the speakers in every case seem to find the idea of "making" a new god unproblematic—it is just something that people do when an occasion momentous enough calls for it. Of course, in "making a god," they would also form new "beliefs" about that deity. So if making the god was a voluntary choice, as it appears it was, forming those "beliefs" most likely was too. Relatedly, appreciators of the Old Testament will realize that phrases like "make a god" are not unique to Nigeria. In Exodus 32, while Moses is on the mountain, the Israelites beg Aaron to "make us gods who will go before us," which Aaron then did by fashioning the golden calf. Creativity in religious "belief" formation sometimes works like this: people choose to make one or more representations of a deity, then they choose to "believe" that the entity represented is real and has supernatural powers. In such cases, the worshippers do *not* come across antecedent evidence from which they infer that some deity exists already; rather, feeling the need for a divinity to worship, they creatively "make a god"—and religious credences to go with it—which is a choice.

Returning to the United States, we find evidence of creativity in the origins of three religious movements outside mainstream Christianity. The key work comes from religious studies scholar Ann Taves, who examines early documents from the three movements: Mormonism, Alcoholics Anonymous, and A Course in Miracles (a religious self-help program).[16] In each case, she finds that one founder of the respective movement—Joseph Smith, Bill Wilson, or Helen Schucman—experienced "revelatory events": they had experiences of visitations from otherworldly presences. A simplistic portrayal would have it that these founders just "believed" what they experienced and then convinced others. But according to Taves's detailed scholarship, the original revelatory experience for the founder was nebulous and open to many interpretations. In all three cases, *a small group of devoted followers* formed around the person who had the initial vision, and they worked out their ensuing mythology and official doctrines over the course of many long discussions and collaborative efforts. Furthermore, the resulting official doctrines and mythology were influenced by various motivations of individual small-group members, as well as the need to reach a wider audience. In no case did the visionary founder just have a clear vision and write it down. Rather, a cluster of hazy yet inspiring spiritual experiences in one individual launched a small group's negotiations over what the meaning of those experiences would be. This is *group creativity* in the formation of religious credences. My point is that such group creativity could hardly proceed if the religious credences they generated had not been under a fair degree of voluntary control.

Some form of creativity—call it recombinatorial creativity—is also present in the next phenomenon of interest.

Syncretism. Syncretism occurs when people (re)combine elements of more than one religion into an amalgam religion. Examples range from syncretism between Hinduism and Buddhism in the Khmer Empire in the twelfth-century CE under Jayavarmam VII to the blending of ancestor worship and Christianity throughout Sub-Saharan Africa. And let's not forget that Christmas—the Christian celebration of Jesus's birth—incorporates festival traditions, such as the decorated "Christmas" tree, of Saturnalia—the Roman celebration of the god Saturn. It is not a stretch to say that any major religion that is currently practiced has at least some syncretism in its history. But let's focus on one example. Boyer describes the mixing of Hindu and Muslim religious elements in Java in the ritual religious meal called a *slametan*. In this event, multiple gods, saints, and ancestors are invoked in a "polytheistic jumble." But Boyer makes clear that this creative mixing of elements is due not to confusion but rather to the need to navigate the political differences and pressures that separate distinct religious communities:

> In such a situation, it is clear that one must "*choose*" a religious affiliation in the sense of joining a particular coalition. By joining the Muslims you identify with a particular faction in a particular political context. . . . By joining the Hindu "camp" you are joining another coalition. Now people are in fact rather reluctant, for reasons that their history explains all too well, to be formally identified as members of this or that coalition. This is precisely because they perceive the risks associated with this kind of coalitional game.[17]

The solution to these competing pressures is to adopt a mixed religion, which enables one to move in *both* political contexts. Thus, adopting syncretic religious attitudes is a voluntary choice one makes to avoid downstream social and political difficulties—or to realize certain social incentives. This brings us to the next topic.

Incentives. Unlike with factual beliefs, people can adopt new religious credences on the basis of incentives. Many incentives have played this role over time, but those I discuss here are (i) political pressures and threats to personal safety and (ii) romantic opportunities.

Forcible conversions, which exemplify (i), are common in the history of Christianity. In this scenario, the alternatives for the would-be convert are typically (a) death and/or torture or (b) conversion. One historical example of mass forced conversion is Charlemagne's conquest of Saxony (now Northern Germany), where many inhabitants up until his invasion in 772 CE still practiced Germanic Paganism (a polytheistic tradition related to Norse mythology). Charlemagne used brutality both to suppress uprisings and to bring about religious conversion. According to his chroniclers, he had forty-five hundred Saxons beheaded in Verden in 782 for their uprising. Later, Charlemagne instituted forced mass baptisms. It was a turbulent transition that lasted about thirty years, but eventually, all of Saxony was Christian.[18]

The individual psychodynamics of these forced conversions must have been various, with many people merely pretending to "believe" and others "sincerely believing" (we'll turn to that "sincerely" at the end of Chapter 6). So it is impossible to say what was going on in any individual case. But we can extract two important points:

1. Given that all of Saxony did become a stably Christian region, it is fair to assume that at least *some* of the conversions were sincere (meaning the converts adopted the Christian religious credences sincerely).
2. Charlemagne thought that making people change their religious "beliefs" through force was a feasible enterprise.

Thus—point 1—when forced to choose between death and Christian "belief," many Saxons apparently did *choose* Christian "belief." And that choosing implies that those "beliefs" were voluntary in the senses described in Chapter 2.

Point 2 is more subtle. It shows that Charlemagne understood the voluntariness of religious "belief": that it was possible to induce the "belief," say, *that Jesus is God* by getting people to choose between that "belief" and death. Here a contrast is interesting. In addition to forced conversions of religious "belief," Charlemagne instituted schools throughout his empire. But why? Well, one does not change ordinary factual beliefs about, say, geography through death threats and forced conversion. It would be bizarre to attempt to persuade people that the Nile is in Egypt by threatening to chop off their heads. As noted in Chapter 2, one changes factual beliefs by informing (hence the schools) or deceiving, not by incentivizing. Charlemagne evidently understood the difference.[19]

Let's now turn from that gruesome history to a more cheerful incentive that has led to religious credence formation time and again: (ii) romance. Examples could fill a book, but I'll confine myself to a few notable political ones.

In 496 CE, King Clovis I, who united the Franks, converted to Catholicism at the instigation of his wife, Clotilde. This was a key event in the rise of Catholicism in the region that would become the Holy Roman Empire.

In 911 CE, Rollo the Viking converted to Christianity as part of a deal with the Frankish king Charles the Simple. That deal, notably, included marriage to Charles's Christian daughter.

In the 1520s, a young Polish woman, Rolexana, was sold into slavery and eventually became the most favored concubine of Sultan Suleiman I in Istanbul. She later became his wife and ruled alongside him, which wouldn't have been possible had she not converted to Islam.

In 2007, former British prime minister Tony Blair, whose wife is Catholic, converted to Catholicism. Not surprisingly, he waited until he was no longer prime minister to convert away from England's official Anglican Church. He and his Catholic wife had been married for a long time, and thus he may have had Catholic leanings going back years before his official conversion. Evidently, his internal mental scale of costs and benefits tipped once he was no longer in office.[20]

One might respond that in such conversions, converts merely adopt the practices without the "beliefs." But marriage- or romance-based conversion is so familiar that it would be implausible to think that *all* such conversions were merely apparent. We should conclude that there have been (and continue to be) many *genuine* conversions for the sake of having a romantic partner who would be unavailable without conversion. That, in turn, would not be psychologically possible if those "beliefs" were involuntary in the way that factual beliefs are.

Now recall the quotation from one of Tanya Luhrmann's informants from the last chapter: "There is always a choice to believe what it is."[21] This subsection contends that there is evidence of people's *choosing* religious "beliefs" at many points in history and across many cultures (not just WEIRD ones). People form new religious credences creatively. They often choose new credences syncretically and for political reasons. And they have time and again voluntarily formed them in response to incentives like physical punishment or romance. Such choices were often difficult, no doubt, but the fact that they were *choices* at all points to the voluntariness of religious credence across time and place.

We must not forget such choices are made based on something that one takes to be the case; that is, based on beliefs in some sense. And when one converts to a religion—forms new religious credences—to attain romance, to avoid punishment, or to achieve political gain, that is in part because one factually believes a certain romantic union is conditional on this, that punishment for nonconversion is imminent, or that new credences can help one navigate a rocky political landscape. Here again, we see the fundamentality of factual beliefs: not only are they involuntary; they are also the cognitive input into the choice to form new religious credences when one chooses to do so.[22]

2.2. Religious Compartmentalization across Cultures

In the last chapter, we saw how Vineyard members admit that they "forget" about God when it's not Sunday and how they only act as if demons exist when other Vineyard members are around—that is, in sacred practical settings. A general One-Map Theorist would argue that Vineyard members are idiosyncratic in that regard, that in other places, people's acting as if supernatural beliefs are true is constant across time and place. But much from anthropology and crosscultural psychology shows otherwise.

Rita Astuti, the world's foremost anthropologist of the Vezo tribe in Madagascar, has long explored their "beliefs" about the afterlife. The Vezo religion—using "religion" broadly—is a form of ancestor worship. Members of the tribe "believe" in some sense that their deceased ancestors still live as spirits and hold sway in the fortunes of the living, so they must be shown reverence. But Astuti notes that whether Vezo individuals act as if their deceased ancestors continue to live is both (i) dependent on context—with ritual settings eliciting more behaviors that treat ancestors as living—and (ii) subject to breaks that reveal thinking according to which the ancestors simply are no more. She writes:

> When the head of my adoptive [Vezo] family addresses the dead, he always ends his whispered monologues by stating loud and clearly: "It's over, and there is *not* going to be a reply!" Every time, the people around him laugh at the joke as they get up to stretch their legs and drink what is left of the rum. But what exactly is the joke? The humour, I suppose, lies in imagining what would happen if one were to expect a reply from dead people, as one does when one talks with living interlocutors: one would wait, and wait, and wait! In other words, people laugh because, as the ritual setting draws to a close, they shift out of a frame of mind that has sustained the one-way conversation with the dead and they come to recognize the slight absurdity of what they are doing. Indeed, my father's joke is probably intended to encourage and mark that shift, as he brackets off the always potentially dangerous one-way conversation with his dead forebears from ordinary two-way conversations with his living friends and relatives. The point I wish to stress is that it takes just a simple joke to break the spell and to call up one's knowledge that the dead can't hear or see or feel cold or, indeed, give a reply.[23]

This is compartmentalization, whereby a joke about how the dead can't respond marks the end of the ritual compartment/practical setting. To use Luhrmann's terms, Astuti describes a transition from a faith frame to an everyday frame, with credences compartmentalized to the former. For all this to be possible, there must be two cognitive maps: one in which the ancestors are alive and can hear, and another in which they are completely gone. The adoptive father's joke is irreverent according to the credence map, on which the ancestors are still listening, but obviously true according to the mundane map. The combination of irreverence and obviousness makes it perfect for a funny transition between practical settings.

Astuti is clearly describing a two-map cognitive structure. Reflecting on her two distinct accounts of how the Vezo relate to the dead, she writes:

> One account delivers the answer that the deceased will continue to want, to feel cold and hungry, and to judge the conduct of living relatives; the other account delivers the answer that after death the person ceases to be a sentient being. In other words, the two accounts manifestly contradict each other.[24]

We thus see the differentiation under discussion: factual beliefs represent the deceased as unliving, while religious credences, as secondary cognitive attitudes, represent them as alive—where the latter attitude type is compartmentalized to ritual settings but the former is the default and hence setting-independent.

A skeptic of this interpretation might say that *both* cognitive layers are practical-setting-dependent: that the credence layer that represents the ancestors as alive is indeed practical-setting-dependent, but the layer that represents them as dead and gone is setting dependent *too* since it turns off in the ritual setting.[25] Hence, on this view, there's no deep difference in this respect between religious credence and factual belief. But this objection misses crucial points that have been with us since Chapter 2. Factual beliefs guide action in settings of make-believe (analogously, during ritual settings), *even when* one is acting on secondary attitudes that run contrary to them. If one is pretending that the concrete floor is a soft bed, one still doesn't flop down on it but rather stretches oneself out slowly; the factual belief that the concrete is hard (continual reality tracking) continues to guide behavior. Similarly, at a burial ritual for a deceased ancestor, *one still moves the body and does not expect it to move itself,* despite the ancestor's being "alive" according to the supernatural ideas at play. So the skeptic's objection is misguided: just like in pretend play, factual beliefs continue to shape behavior, even in make-believe or ritual settings. Astuti also notes that young Vezo children seem to develop the idea of death as total annihilation *before* they develop the idea of death as continuation in the form of a spirit. She adds: "This ontogenetic perspective might explain why the early understanding of death as the end of all sentient life continues to act as a *default,* a default that can only be successfully challenged and overcome in certain limited contexts."[26] (And let us not forget that *something* in people's minds must keep track of what setting one is in—ritual or nonritual?—and the mental state that keeps track of what setting one is in must not be setting-dependent, on pain of regress; that mental state, then, is factual belief, which Michael Bratman calls (using language that mirrors Astuti's) "the default cognitive background."[27])

Experimental evidence further supports Astuti's observations. In one series of studies, Astuti teamed up with Paul Harris to explore differences in Vezo thought and talk about the deceased that depend on whether a tribe member is in a *naturalistic* or a *ritual* context. Astuti and Harris provided Vezo participants with a burial ritual narrative and asked them in that context about the physical and psychological properties of someone who had died. Some questions were related to the body: Do his eyes work? Does his heart beat? Others were mentalistic: Does he miss his children? Does he know his wife's name? They asked other participants the same questions about deceased ancestors described in naturalistic settings (no rituals mentioned). The result was that the Vezo were much more likely to attribute psychological properties (*seeing, thinking,* etc.) to deceased ancestors in a religious-ritual context than in a naturalistic context, which led Astuti and Harris to this striking conclusion:

> Vezo do not believe in the existence and power of the ancestors in the abstract, but they believe in them when their attention is on tombs that have to be built, on dreams that have to be interpreted, and on illnesses that have to be explained and resolved. In other contexts, death is represented as total annihilation, and in these contexts it would be misleading to insist that Vezo believe in the existence of ancestral spirits.[28]

So the experiment yielded the same conclusion as ethnographic observation: being in the ritual-religious "context" toggles Vezo minds to using sacred cognitive attitudes concerning the psychological powers of deceased ancestors, which largely deactivate outside that setting. And the Vezo are not anomalous in their compartmentalization of afterlife credences. The research just mentioned is the sequel to Harris's research with Marta Giménez, which found a similar effect of "context" among Catholic Spanish children, who were also more likely to attribute psychological properties to the deceased when probed in a religious setting. The same effect has also emerged in places as diverse as Austin, Texas, and the Melanesian island of Vanuatu.[29]

So afterlife "beliefs" are commonly compartmentalized. What might have led so many other researchers to miss this point? Here, an observation from Evans-Pritchard's 1965 book *Theories of Primitive Religion* is illuminating. ("Primitive," as he clarifies, was during that period of anthropology an unfortunate technical term for preindustrial societies organized in small communities.) The book is an evidence-based roast of prominent nineteenth- and twentieth-century theories of the origins of religion. And one of Evans-Pritchard's main complaints is that many of those theories were built on badly biased observations.

> What travellers liked to put on paper was what most struck them as curious, crude, and sensational. Magic, barbaric religious rites, superstitious beliefs, took precedence over the daily empirical, humdrum routines which comprise nine-tenths of the life of primitive man and are his chief interest and concern: his hunting and fishing and collecting of roots and fruits, his cultivating and herding, his building, his fashioning of tools and weapons, and in general his occupation in his daily affairs, domestic and public. These were not allotted the space they fill, in both time and importance, in the lives of those whose way of life was being described. Consequently, by giving undue attention to what they regarded as curious superstitions, the occult and mysterious, observers tended to paint a picture in which the mystical . . . took up a far greater portion of the canvas than it has in the lives of primitive peoples, so that the empirical, the ordinary, the common-sense, the workaday world seemed only to have secondary importance.[30]

Two points here are relevant for present purposes. First, "humdrum routines" like hunting and gathering constitute "nine-tenths of the life" of those in preindustrial societies, and "religious rites" and "superstitious beliefs"—the "mystical" and "the occult and mysterious"—occupy a far smaller portion. This observation coheres with the idea that the attitudes driving many of the magical practices and religious rites are compartmentalized, while factual beliefs ("the empirical, the ordinary, the common-sense, the workaday world") are practical-setting-*in*dependent. Second, a caution: if you form a theory only from observations of a society's sensational and mysterious practices, you will get a distorted picture of their mental lives.[31] To the present point, if you focus only on situations in which ancestor spirits are being discussed, it will seem that your informants simply factually believe in the continued afterlife of ancestors. But that impression would result from biased observation, which fails to document what things are like when the religious setting ends ("and there is *not* going to be a reply!").

Compartmentalization of supernatural ideas doesn't just occur with afterlife "beliefs." Recall the Legare and Gelman studies of explanatory coexistence when it comes to Sesotho thought about HIV/AIDS. Not only *is* there explanatory coexistence; the studies also describe a difference in compartmentalization: "bewitchment explanations are *contextually specific* in ways that biological explanations are less bound by." And: "biological explanations are the default explanatory framework." Relatedly, their Study 3 finds that participants appealed to biological explanations far more often than bewitchment explanations. Finally, there is an improvisatory character to how participants reconcile bewitchment and biological explanations (note this also illustrates voluntariness): "participants may be creating these solutions spontaneously in the course of the experiment."[32] All of this makes sense in light of a Two-Map Theory: people have practical-setting-*in*dependent factual beliefs about biological causes of AIDS; they have "contextually specific" (practical-setting-dependent) religious credences about bewitchment; the two maps are not confused.

Religious compartmentalization is also evident in "prosocial" or moral behaviors. You will recall that religious credence, as a *cognitive* attitude, ranges mainly over descriptive contents. But since descriptive ideas often bear on moral choices, the practical setting dependence of certain kinds of moral choice reveals the compartmentalization of related religious credences.

To give a specific example of this research, behavioral economist Erik Duhaime investigated the relationship between religious cues and altruism among Muslim shopkeepers in the Medina district of Marrakesh. He presented shopkeepers with an economic game with three options: choosing option 1 garnered 20 dirhams for the participant (a dirham at that time was worth about 11 cents US); option 2 garnered 10 dirhams for the participant, while 30 would go to charity; and option 3 gave 60 dirhams to charity and none to the participant. So, from a selfish perspective, option 1 > option 2 > option 3. From an altruistic perspective (and from the perspective of total amount given out): option 3 > option 2 > option 1. Duhaime's study tested whether hearing *the call to prayer*, which sounds at specific times throughout the day in the Medina, would affect the option participants chose. And it did: 100 percent of participants who made their choice within twenty minutes of the call to prayer picked option 3 (the most charitable one), whereas outside that window of time, the number was 50 percent. The call to prayer, then, is a *cue* that activates a set of religious representations that lead to charitable giving but, in a significant portion of the population, lie dormant outside it.[33]

To return to the American Christian context, two illuminating papers reveal a "Sunday Effect" on morally salient behaviors.

In one, economist Benjamin Edelman discusses markets for online pornography and considers several factors that could affect rates of purchase: location, political leaning, age, socioeconomic status, religion, and others. His main conclusion is that pornography purchases are widely distributed and occur at not-too-different rates across regions and demographics. But there are some interesting differences. It may be surprising to learn, for example, that the state with the fewest pornography subscriptions per capita was West Virginia, while the state with the most was Utah. And Edelman's findings on pornography purchases in regions where the most people go to church (that is, the "Bible Belt") are especially noteworthy for thinking about compartmentalization.

> In regions where more people report regularly attending religious services . . . overall subscription rates are not statistically significantly different from subscriptions elsewhere ($p = 0.848$). However, in such regions, a statistically significantly smaller proportion of subscriptions begin on Sunday, compared with other regions . . . on the whole, those who attend religious services shift their consumption of adult entertainment to other days of the week, despite on average consuming the same amount of adult entertainment as others.[34]

In the other paper, Deepak Malhotra finds a Sunday Effect for charitable giving. Malhotra collaborated with a US-based charity firm that hosts auctions in which people bid on a charity and then can "rebid" in case they've been outbid by someone else. Malhotra wanted to see whether the day of the week would make a difference in rates of rebidding. And it did—among people who identified Sunday as their day of worship.

> On Sundays, whereas religious bidders were 40% likely to re-bid in response to an appeal to charity, non-religious bidders were only 11.8% likely to re-bid in response to such appeals. Notably, on other days of the week, re-bidding in response to charity appeals was almost identical among religious (25%) and non-religious bidders (27%), strongly suggesting that religious individuals are not more pro-social in general; they respond to appeals for help more so than non-religious individuals only when their religion is salient to them.[35]

In terms of altruism, the religious and the nonreligious *look no different* when it's not Sunday. So not only does Sunday activate charity-related religious credences; when it's not Sunday, being Christian has little to no effect. As psychologist Azim Shariff said to me in conversation, "Religious people are people first and religious second."

In sum, ample evidence points to the practical setting dependence of religious credences across cultures. People's afterlife "beliefs," "beliefs" about supernatural causes of illness, and "beliefs" about divine commands are compartmentalized in diverse populations around the globe. Such compartmentalization is all the more striking when we recall, with Evans-Pritchard's later work, that the "empirical" and the "humdrum"—in short, the realm of factual beliefs—occupy "nine-tenths" of people's life activity.

2.3. On Lack of Cognitive Governance

Let's briefly rehash cognitive governance. Suppose the supermarket Cleo goes to (now she's an adult) is usually open until 11:00 p.m. But one day, Kevin, who knows she needs to pick up dog food, informs her the store is closing at 8:00 p.m. Cleo concludes that she can't make it in time since she gets off work at 8:30 p.m.

But why? Why not think that the supermarket will reopen at 8:15 p.m. and stay open the rest of the night? What gets her from the belief that it will close at 8:00 p.m. to the belief that it will *still* be closed at 8:30 p.m.? That further factual belief does not follow by logic alone.

The answer is that other factual beliefs supply the background information. The general factual belief *that once supermarkets close they don't reopen until morning* played the needed role in governing Cleo's inference. And this is cognitive governance: a class of attitudes X governs class Y when it supplies background information for use in inferences from one cognitive attitude in class Y to others in class Y (where Y and X may be the same or different classes). In this example, factual beliefs govern themselves. But recall that they also govern imaginings, suppositions, acceptances in a context, and other secondary cognitive attitudes. As we saw in Chapter 2, this relation is asymmetric: factual beliefs cognitively govern other cognitive attitudes but not vice versa.

The question now is whether and to what extent religious "beliefs" exemplify such governance. We saw that many Vineyard "beliefs" do not do this—at least not in a widespread way. Let's expand our investigation crossculturally.

On this matter, some passages from Mark Chaves's article "Rain Dances in the Dry Season: Overcoming the Religious Congruence Fallacy" are revealing.

The religious congruence fallacy occurs when a theorist of religion reasons as if the contents of people's religious "beliefs" can, based on seemingly logical extrapolation, secure straightforward predictions about how the "believers" will think and behave. Chaves's title points to a great example: anthropologist Meyer Fortes once asked a rainmaker in a traditional society to perform a rainmaking ceremony during the dry season, which the rainmaker found to be an absurd suggestion. The anthropologist had reasoned (sensibly, it would seem) as though the rainmaker's "belief" that the ceremony would be effective at bringing rain would lead the rainmaker to infer that this could be done whenever it would be useful to have rain. The rainmaker scoffed at the inference. I would put the point like this: the rainmaker had general background knowledge, encoded in factual beliefs, about when rains arrive and when they don't, and his religious credences about being able to summon rain through ceremony did not alter the fabric of that background knowledge. The congruence fallacy was instantiated in Fortes's assumption that the rainmaker's religious "belief" would alter that fabric.[36]

Consider these passages from the same paper:

> One way to see incongruence clearly is to examine the most instrumental-looking ritual and religious action. The key observation is that instrumental-looking ritual and religious action usually supplements practical action, *even when congruence would lead us to expect it to replace practical action.* Ludwig Wittgenstein articulated this point when he commented on James Frazer's *The Golden Bough,* which exudes congruence by assuming, for example, that people believe that stabbing an enemy's effigy before battle guarantees military success, or that morning rituals make the sun rise. In considering these examples, Wittgenstein pointed out that the same person who stabs an effigy also carefully crafts and sharpens his weapons. About dawn rituals he said: "The rites of dawn are celebrated by the people, but not in the night; rather there they simply burn lamps." . . .

Many people believe in divine healing and actively seek it, for example, and prayers and anointing for healing, and testimony about divine healing, are common at Pentecostal and other worship services. But very few people seek divine healing *instead of* medical treatment—unless, of course, they don't have access to quality medical care. . . . Divine healing testimonies often contain the exclamation—"The doctors were amazed!"—again indicating that healing prayers and rituals mainly *supplement* rather than *replace* medical care, just as superstitious athletes can believe that rituals or talismans improve their play while they also train and practice incessantly. People all the time pray for health or wealth or victory in battle, and such instrumental-looking religious action may look like dancing for rain in the dry season. But the fact that such action almost always supplements rather than replaces practical action shows otherwise.[37]

Two things stand out here.

First, prayers, rites, and other rituals are almost always *supplemental* to instrumental action (or to events that will happen anyway), just like in Vineyard petitionary prayer. When people stab the effigy though still sharpening spears, perform the rite of dawn only when the sun is about to rise, or pray for a cure when going to the doctor or taking medicine, they have a stable background picture of how the world works that is largely the same as that of a nonreligious person: sharp spears are more effective, the sun appears in the morning, and certain medical treatments help cure certain diseases. This stable background remains unchanged by the respective religious "beliefs." That's why religious action that *looks* instrumental is usually really supplementary to actions one would perform whether or not they had the relevant religious "beliefs." Otherwise put, even when people "believe" that gods can give victory to those with dull spears or that God can cure an illness, the fact that they do not act as if they *anticipate* these things reveals that their related factual beliefs are *not* governed by their religious credences.[38] Rather, their factual beliefs leave certain possibilities open, and the outcomes people typically pray for fall among the open undecided possibilities.[39]

Second, Chaves writes as though the supplemental nature of religious "instrumental" action is obvious to many scholars who have studied religious communities extensively—as if the point is commonplace. The corollary point should be just as obvious: religious credences do *not* govern factual beliefs. If religious credences governed factual beliefs, we would expect inferences like this to occur:

RELIGIOUS CREDENCE: *performing the rite of dawn makes the sun come up*
FACTUAL BELIEF: *it is right now the middle of the night*
*FACTUAL BELIEF: *performing the rite of dawn right now will make the sun come up in the middle of the night*

In this triad, the asterisk marks the factual belief that people are unlikely to form, even though the first two mental states in the series are common enough. That such an inference is unusual at best attests to the existence of religious credences (about the causal efficacy of the rite) that lack widespread cognitive governance. A worthwhile question is this: If the performers of the rite *didn't* have stable factual beliefs about when the sun appears, how would they know to perform the rite always at just the right time?

Another indication of the lack of widespread cognitive governance on the part of religious credences is *inferential curtailment.* This occurs when an individual or group either fails to draw or avoids drawing obvious inferences from professed "beliefs" (and it is unlikely that this is due to forgetting, oversight, or processing difficulty). According to Scott Atran, who did anthropological fieldwork among the Mayan-speaking Itza', his informants professed that humans transform into animals. Yet they never worried that, when eating animal meat, they might be eating a human. Atran wryly notes that they "should suspect someone eating a porkchop might be a cannibal."[40] But they don't, so their inferences from "beliefs" about humans turning into animals are curtailed.[41] Curtailment, furthermore, is common in various religious contexts. How many professed biblical literalists, for example, actually accept the inference that the children of Adam and Eve were incestuous? Congruence would lead us to expect that inference, but my impression is that most just avoid the issue. The inference is curtailed.[42]

A further manifestation of lack of widespread governance is *theological correctness.* Psychologists Justin Barrett and Frank Keil, who coined that term, tested people from a range of religions to see whether their theological "beliefs" ("beliefs" that encode official church doctrines) would guide

memory of and inferences about stories involving a deity. Their participants came from diverse religions—Bahaism, Buddhism, Catholicism,
Protestantism, and Judaism—and all endorsed the "omni" properties of
God: God is omnipotent, omniscient, and omnipresent. Barrett and Keil
hypothesized, however, that such theologically correct beliefs would not
influence recall of a story. They gave participants a fictional story about
a people in need of help in two separate places on earth to see whether
the participants would recall God as helping *sequentially* or *simultaneously*,
though the vignette left it open which one it was. One story was as follows:

> A boy was swimming alone in a swift and rocky river. The boy got his left
> leg caught in between two large, gray rocks and couldn't get out. Branches
> of trees kept bumping into him as they hurried past. He thought he was
> going to drown and so he began to struggle and pray. Though God was
> answering another prayer in another part of the world when the boy
> started praying, before long God responded by pushing one of the rocks so
> the boy could get out. The boy struggled to the river bank and fell over
> exhausted.[43]

One interpretation of the text is that God kept on helping elsewhere while
helping the boy. Barrett and Keil found, however, that most participants
recalled the story with God helping *sequentially*; God finished helping in
one place and *then* helped the boy. But if God is omnipresent and omnipotent, He should be equally likely to help simultaneously—if not *more*
likely (after all, presumably, He is perpetually performing countless
helping acts simultaneously). The fact that people almost never remember
the story that way, Barrett argues, shows that they have an *intuitive* God
conception, in addition to their theologically correct one, where the intuitive one portrays God as an anthropomorphic agent who is limited in
space and time. This intuitive conception, rather than the theologically
correct "belief," guides their reconstruction of the story.[44]

Theologically correct credences are largely inert when it comes to guiding inference, so they lack widespread cognitive governance. But then,
should we consider the "intuitive" conceptions of God that Barrett and
Keil discuss to be factual beliefs since they *do* seem to exhibit cognitive
governance (at least in story recall)? That is tempting, but consider a quotation from one of Barrett's informants: "I just kind of always picture God
as like an old man, you know, white hair . . . kind of old, I mean, but

I know that's not true."[45] Here, the informant, who is representative of many "believers," explicitly disavows the intuitive God conception, presumably out of awareness that it sits ill with the theologically correct one. And this disavowal, if we take it as genuine, takes the intuitive conception out of contention for being a factual belief. In doctrinal contexts, the intuitive conception would clearly *not* govern certain inferences. So we are left with the result that *neither* people's intuitive conceptions of God *nor* their theologically correct conceptions of God have widespread cognitive governance.[46]

The picture emerging from these examples is that there is a curious selectivity to the inferences that are drawn from religious credences. Sometimes, inferences are drawn that logic and background information would lead one to expect, but sometimes not (and it is unlikely that the lack of inference is due to processing error). Think how strange it would be if Cleo factually believed the store closed at 8:00 p.m. but did not infer that it would still be closed at 8:30 p.m.—without any additional explanation. But analogous curtailment happens pervasively with religious credences, as we just saw. This is evidence of a lack of widespread cognitive governance. We should note that it raises a further question: What *determines* *when* inferences will be drawn from religious credences and when they won't? We'll come to that in Chapter 6. We can now turn to the fourth feature of factual beliefs: evidential vulnerability.

2.4. How Religious Credences Lack Evidential Vulnerability

If you ask religious "believers" for *evidence* of their stories, doctrines, and supernatural ontologies, some will produce what sounds like answers. Others will say you're missing the point; it's about faith. My sense is that producing "evidence" for religious claims is more prevalent in Western contexts than elsewhere. But that is no small thing, and it occurs outside the West as well: I once had a guide in Cambodia who, in addition to Buddhism, practiced animism, and he was eager to tell me what he took as evidence of healing spirits.[47] In any case, theists often produce arguments that treat the appearance of design in nature as "evidence" that God exists. Apologists talk about religious texts as "historical documents." Systematic theology attempts reasoning that aims at being metaphysical proof. And in an everyday way, many devout people cite unlikely but beneficial coincidences as "evidence" that God cares for them. Do such evidence-citing behaviors support the idea that religious "beliefs" respond to evidence?

Such a conclusion would lose track of the dialectic.

Recall from Chapter 2 that *evidential vulnerability* was one property that helps distinguish factual belief from fictional imagining, supposition, practical assumptions, assumptions for the sake of argument, hypothesis, and any other secondary cognitive attitudes (recall Hume's Desideratum). That was the point of the principles I laid out, including this: *factual beliefs respond to evidence.* But that principle was only useful for capturing the needed distinction if it was sharpened in the right way.

Importantly, factual beliefs are not the only cognitive attitudes for which people cite evidence. I often marshal evidence for my hypotheses even and perhaps especially when I don't (yet) factually believe their contents. That might sound like confirmation bias—and maybe it is—but the point is that the mere tendency to *gather* evidence for the contents of a mental state does not help distinguish that state from factual belief; it might even be more likely for states that are *not* factual beliefs. Alternately, if I assume that *p* for the sake of argument, I will often produce evidence for *p* in the course of discussion. Even things one fictionally imagines can have *some* evidence in their favor, which one might cite in the right sort of game. So, again, people's tendencies to cite evidence for their religious "beliefs" do not distinguish them from other secondary cognitive attitudes in this regard. Such tendencies are neither here nor there with respect to the major question on the table.

Evidential vulnerability, however, is a <u>distinctive</u> property of factual belief. It is an *involuntary proneness to being extinguished* by contrary evidence. It is the extinction condition that is distinctive. Thus, the principle that *factual beliefs respond to evidence,* suitably sharpened, means that if a person who holds a factual belief cognizes strong evidence contrary to it (perceptual or otherwise), it is prone to being extinguished—whether the person likes it or not. Thus, if we wish to find out whether a certain class of mental states is evidentially vulnerable, we should look for examples not of positive evidence-citing behavior but of how those attitudes respond in the face of strong contrary evidence. In particular, if those mental states persist in people's minds, even when they have cognized strong evidence that they are false, that persistence is evidence against the idea that that class of states is evidentially vulnerable.

Now consider doomsday cults. As has been famously documented, cults that have predicted doomsday on a certain day persist and even increase *after* that day has passed.[48] Their "belief" system survives decisive evidence against the focal "belief" in it. Comparable factual beliefs aren't like this. Before January 1, 2000, for example, many people predicted that the world's computer systems would go haywire on that day due to the "Y2K problem."[49] But a combination of good fortune and extensive work by diligent programmers ensured that the Y2K meltdown never occurred. So when January 1 arrived with no major problems, most people's Y2K factual beliefs were promptly extinguished—*poof!*—by the evidence (unlike doomsday religious credences, which weren't).[50]

There are many other examples of evidentially disconfirmed religious "beliefs" that persist. Many Mormons still have the religious credence that American Indians descended from the lost ten tribes of Israel. In particular, the Book of Mormon says American Indians descended from the cursed (and hence darker-skinned) Lamanites, who make up one of the two offshoot groups of the lost Israelite tribes (the other being the lighter-skinned Nephites). Such "beliefs" are astonishing since, in addition to their racism, the historical and archaeological evidence shows that the ancestors of American Indians crossed over from Asia via the Bering Sea.

Nevertheless, one intellectually curious Mormon put the idea of Israelite ancestry to a DNA test in 2002. Thomas Murphy, a Mormon and anthropology professor at Edmonds College in the Seattle area[51] reasoned that if the Mormon "belief" about American Indian ancestry were true, DNA evidence should confirm it—or refute it. But when he gathered the evidence, it did not confirm the "belief"; rather, it pointed to Asiatic origins only, like most other evidence. What effect did that seemingly decisive disconfirmation have on church officials? Instead of rejecting their religious credences, the local Mormon stake summoned Murphy for excommunication. On January 12, 2003, the *Seattle Post-Intelligencer* reported as follows:

> In December [of 2002], the local stake of the Church of Jesus Christ of
> Latter-day Saints scheduled a disciplinary council and informed Murphy
> he faced the possibility of excommunication, or expulsion from the church.
> But the president of the stake—a district made up of a number of wards—

indefinitely postponed the council after the debate hit the press and sup-
porters staged rallies across the country.[52]

The stake backed off due to *political* pressure, not evidence, but the threat
of excommunication had been real. The incident has long since blown
over. Murphy is no longer Mormon, but that is due to voluntary departure rather than excommunication. To the present point, however, the
LDS Church officially still affirms the idea that American Indians descended from Israelites, despite all the evidence to the contrary.[53]

One could point to many other religious "beliefs" that persist despite
being out of keeping with known evidence, such as the "belief" of Young
Earth Creationists that the earth is less than ten thousand years old. One
could also point out that toleration of inconsistencies among religious
"beliefs" is common in various traditions—from biblical inerrantists to
Thai Buddhists who maintain inconsistent positions concerning the existence of forest spirits.[54] But an even more striking indicator of evidential
invulnerability in religious "beliefs" is that people maintain them despite
seriously *doubting* them (just like in the Vineyard). In his recent book *The
Lies That Bind,* Anthony Appiah describes his mother's process of being
confirmed as an Anglican:

> [My mother] told me that when she was preparing for confirmation, the
> ceremony that marks the transition to full adult membership of the Angli-
> can Church, she mentioned to her father that she was having difficulty
> with some of the Thirty-Nine Articles of faith that have defined the dis-
> tinctive traditions of the Anglican Church since the reign of Elizabeth I.
> "Well," Grandfather said, "I have a friend who can help you with that."
> That friend was William Temple, then the Archbishop of York, and later
> the Archbishop of Canterbury, spiritual head of the Anglican Church. My
> mother went to see him. As they scrutinized the Articles, every time my
> mother said that something was hard to believe, the archbishop agreed
> with her. "Yes, that *is* hard to believe," he would say. She went home and
> told her father that if you could be an archbishop with her doubts, you
> could surely be an ordinary Anglican.[55]

The point is clear. The archbishop himself clung to his Anglican "beliefs"
despite having persistent doubts about them. His having doubts shows
that he had *registered evidence* contrary to those "beliefs" (else why doubt

them?), but he clung to his religious credences in the Thirty-Nine Articles nonetheless. These tutoring sessions, then, amounted not to a presentation of evidence for the articles but to a training in what attitude to have toward them: "believe" them, despite doubt and lack of evidence. In other words: have religious credence.

"Belief" in the face of doubt—or at least lower epistemic confidence—is not just a Vineyard or Anglican phenomenon. Crosscultural studies show that even in religious societies, such as Iran or Spain, people express higher levels of confidence in the existence of unobservable scientific entities, such as germs or oxygen, than in the existence of supernatural entities, such as God or angels.[56] This difference already begins to appear around the ages of ten to twelve. Such findings have also emerged in places as diverse as the United States and China. Considering this difference in confidence and relating it to verbal justifications given by the young experimental participants, Sylvia Guerrero and her colleagues summarize what they discovered among religious schoolchildren in Spain:

> Summing up, there are several grounds for concluding that older children do not conceive of invisible, religious entities and invisible, scientific entities in the same way. In particular, children do embed scientific entities into a causal sequence. For example, "We can breathe with oxygen," "Germs exist—because of them we get diseases." More generally, it is feasible that children are confident about the existence of germs and oxygen, not simply because they hear other people endorsing their existence but because they understand and accept larger causal narratives in which germs and oxygen play a central role. Thus, they have some understanding of the way that germs can be transmitted, the way that they multiply, and their role in the spread of contagious disease. Similarly, they recognize the role of oxygen in the processes of respiration and combustion. By implication, children firmly accept the existence of unobservable entities provided they play a plausible explanatory role with respect to readily observable phenomena such as illness and breathing.[57]

In short, factually believing invisible scientific posits is not merely a matter of taking people's word for it; such posits are tested and retested every time someone as young as ten plugs in a toaster or blows on coals. It is no surprise, then, that lower confidence levels exist for entities who must be taken as given on the basis of testimony alone and whose effects (like answered prayers) are unreliable at best. Yet despite lower confidence levels, religious credence persists.

This all raises the question of how people psychologically reconcile "beliefs" that don't square with evidence with the common expectation that "beliefs" (of whatever sort) *should* square with evidence. Here, research by psychologists Emily Liquin, Emlen Metz, and Tania Lombrozo is pertinent. They compared how participants evaluated religious "why" questions ("Why is there an afterlife?") to how they evaluated scientific "why" questions ("Why is the center of the earth so hot?") and examined two questions. First, to what extent did participants indicate an explanation was needed? They called this *need for explanation.* Second, to what extent did participants accept "It's a mystery" as a good answer? They called this *mystery acceptance.* Their finding across studies was that participants judged scientific "why" questions to need much more explanation than religious "why" questions, whereas mystery acceptance showed the opposite pattern. Thus, people in their sample tended to treat religious claims as if they did not need to be logically integrated into the fabric of known facts—saying, "It's a mystery" is fine.[58] This differentiates the claims that express religious credences, not just from scientific claims but also from commonsense ones. If you asked someone in your house why the cookies are gone, you would not be satisfied if they said, "It's a mystery."[59]

In sum, there is extensive evidence that many religious "beliefs" among various peoples in various traditions are *not* vulnerable to evidence. Many religious credences survive disconfirmation by the relevant facts. Religious people in diverse places around the globe harbor doubts while still clinging to their "beliefs," which shows that at some level, many of them have registered relevant contrary evidence or the complete lack of evidence. And all this is often internally justified by an outlook that tolerates mystery and denies the need for explanation. It is true that *some* people leave their religions because the evidence disconfirms their former "beliefs." There are people like Thomas Murphy. But as Jon Bialecki points out in relation to the Vineyard, evidence is far from the most common reason for leaving. And as sociologist Romy Sauvayre shows, people's reasons for leaving a cult usually have more to do with a *conflict of values* than with the fact that the cult's "beliefs" are out of keeping with evidence.[60] And that's because religious credence, as a mental state type, is not vulnerable to evidence.

3. CONCLUSION: RELIGIOUS CREDENCE IS NOT JUST WEIRD

It is common in many cultures—not just WEIRD ones—for religious "beliefs" not to have the characteristic properties of factual beliefs. In the commonsense language discussed in the next chapter, religious people may "believe" their doctrines, myths, and supernatural ontologies, but that does not entail that they straightforwardly <u>think</u> they are true; actions and inferences often tell otherwise. In my terms, they religiously creed those things, but that does not mean they factually believe them.

Religious credence is a second layer in a two-map cognitive structure, which helps explain explanatory coexistence globally. It is voluntary and regarded as such, which explains the language of "choice," creativity, syncretism, religious compulsion, and "belief" due to incentives like romance. It is compartmentalized to sacred times and places, which helps explain a wealth of ethnographic observations and systematic behavioral studies. It is lacking in widespread cognitive governance, which makes sense of the lack of religious congruence, to use Chaves's term, as well as inferential curtailment and theological correctness. And it is not vulnerable to evidence in the relevant sense, which helps explain mystery acceptance and the common lack of need for explanation. None of this, I reiterate, implies that religious credence is without serious consequence. Far from it: religious credence shapes everything from marriages to executions and from art to architecture. But all that is consistent with (and even better explained by, to the extent that those cultural forms are *symbolic*) its being a secondary cognitive attitude, one that constitutes group identity and activates sacred values. The last chapter showed that religious credence exists; this chapter showed that it is widespread in time and place.

CHAPTER FIVE

To "Believe" Is Not What You "Think"

1. A DISTINCTION YOU ALREADY KNOW

You might wonder whether religious people—or people in general—are aware of a difference between religious credence and factual belief. After all, it is one thing for there to *be* distinct mental kinds (this has been the focus of my argument thus far); it is another for people *to be aware* that they are distinct (since much psychological processing occurs without awareness).

Is it just people who have read books like this one who are cognizant that factual belief and religious credence are different? Or do neurotypical people more generally (those with normally functioning social cognition) at some level view the respective mental states in different lights? As we saw, Vineyard members think extensively about their religious "beliefs," but do they or others cognize religious credence and factual belief *differently*?[1]

I suggested an answer to this in the Prologue, where I said that I think some part of you intuitively knows the idea that I'm trying to advance. The main aim of this chapter is to substantiate that suggestion by presenting linguistic and psychological evidence that people generally—

across diverse cultures—think and talk about religious cognitive attitudes ("beliefs") differently from factual beliefs. In part, that should not be surprising: the ability to attribute cognitive attitudes is not rare; neurotypical people around the globe begin doing it at a young age (at least by the time they pass the false belief task).[2] What may surprise some is that people also differentiate between religious credence and factual belief.

The following anecdote points at the phenomenon of interest.

One evening in 2016, while I was living in Antwerp, I visited my friend Dirk for drinks and dinner. Dirk was an Anglican minister (now retired), and he was interested to hear my emerging views on belief. I decided to start him off with the sort of example of "belief" one gets in philosophy (in epistemology and philosophy of action, especially). Then I would carve out my position from there. As it happened, there was a giant copper beech in the yard where we were sitting.

So I said: "You, for example, have a <u>belief</u> that there's a copper beech in the yard."

He shot back: "That's not a belief. That's a fact!"

I could glimpse why he said that, but I persisted. I pedantically replied as many philosophers would: yes, it was a fact, but unless he had some internal mental representation of that fact in his mind/brain, he wouldn't be able to act on it or report it. And that internal representation, given a certain attitude attached to it, is what philosophers call a belief. Dirk nodded while looking at me suspiciously. He felt that if *that* is what one means by "belief," then (sure) he had a "belief" that there was a copper beech in the yard. But that was not the sort of "belief" in which he—as an Anglican minister—was interested!

The point of this anecdote is *not* that Dirk had no concept of the mental state that philosophers typically label belief and that I call factual belief. He did; otherwise, he wouldn't so easily have understood my explanation of the mental state when I gave it. The point is rather that he did not primarily associate the *words* "believe" and "belief" with *that* mental state. Evidently, he wanted to use the word "belief" for something different and had assumed that I had planned to do the same, which is why, though he understood, he looked at me suspiciously. This suggests that Dirk was in fact one step ahead of me in the conversation: he was *already* differentiating religious credence and factual belief in some vague way while preferentially associating the word "belief" with religious credence (the one he *took* himself to have been asking about).

I think that many people think and talk like Dirk in this respect.

Broader linguistic evidence for this, which appears in section 2, comes from patterns of difference in how people use "think" versus "believe." Roughly put, people are more likely to use "think" for reporting other people's matter-of-fact cognitive attitudes ("Jane thinks her bike is in the garage"), while they are more likely to use "believe" for reporting other people's religious cognitive attitudes ("Janet believes there is only one God"). And that's not just for English. My colleagues and I have found analogous patterns of differentiating word choice crossculturally: similar results emerge in Ghana (among Fante speakers), in Thailand (among Thai speakers), in China (among Mandarin speakers), and in Vanuatu (among Bislama speakers).

Psychological evidence, presented in section 3, includes the fact (among others) that people apply different norms to religious versus factual beliefs when it comes to (apparent) *disagreements,* as shown in earlier work by Larisa Heiphetz, who went on to collaborate with me on some of the word choice studies just mentioned.

I have three reasons for emphasizing all this before further extending my theory in the coming chapters.

1. The data I present supply an indirect argument for the distinction between factual belief and religious credence: that is, part of the explanation for *why* people talk and think in the differentiating patterns they do is that there is a difference in attitudes that they are tracking.
2. I think that highlighting the empirical data I do will enable readers to connect the technical notions of factual belief and religious credence to intuitions they already have, which will facilitate both better understanding of the theory and better awareness of why those intuitions exist in the first place.
3. The present material provides a natural transition to the next chapter, where I discuss the connection between religious credence and group identity.

According to fellow philosopher Eric Schwitzgebel, the developmental psychologist Alison Gopnik once remarked that one way to tell that someone doesn't believe something is if they *say* they "believe" it.[3] Another way of looking at this chapter is that I'm developing intellectual tools that will enable you to unearth the sense beneath that paradoxical-seeming comment, the explanation for which comes in section 4.

2. "THINK" VERSUS "BELIEVE"

One evening in 2012, my colleague Dan Weiskopf mentioned an observation he'd made in his philosophy of mind classes. He said it's a bit of work to get students to use "believe" in the way philosophers do; most people gravitate to using "think" in that role. For example, the sentence "Fred believes there's beer in the fridge" (which is normal for philosophers) is not quite as idiomatic as "Fred *thinks* there's beer in the fridge." Conversely, it's natural to say, "Hannah *believes* Jesus rose from the dead," but "Hannah thinks Jesus rose from the dead" sounds odd.

The difference between philosophical and lay use seems to be even more striking for the noun "belief." Philosophers (and many cognitive scientists) use that noun for the mental state people are in when they simply think something is the case (in fairness to us, it's useful to have such a noun). So in many philosophical writings, you'll see this awkward usage: "Fred has the <u>belief</u> that there is beer in his refrigerator" (which is often followed by something like "That <u>belief</u> causes and rationalizes his behavior of going to the fridge when he wants beer"). But ask laypeople what some of their "beliefs" are, and you are much more likely to hear about religious, spiritual, and ideological "beliefs"—even faith. You are unlikely to hear about their "belief" that there is beer in the fridge. (If you were to ask a man on the street what some of his "beliefs" were and he replied, "I have the belief that there is beer in my fridge," that answer would be funny precisely *because* (a) its "belief" usage crosses mental state types [religious credence and factual belief], (b) it suggests a sacralizing regard for beer, and (c) in so doing, it implicitly demotes other things that people more typically regard as sacred.)

I immediately saw the relevance of Dan's point to my broader project. An objection I had heard from other philosophers to my distinction between religious credence and factual belief went like this: "But people call their religious beliefs '*beliefs*!'" The implication was that people must be thinking of their own and others' religious beliefs as factual beliefs (albeit ones with religious contents); after all, that's how philosophers typ-ically use the word "belief."[4] But that objection only works on a mistaken assumption that only a philosopher would make: that nonphilosophers, when talking about religious "beliefs," are using that word in the same matter-of-fact belief sense that philosophers often give it. If Dan was right (and if Dirk is a good example), that's not the case, and people's discrimi-nating word choice ("think" versus "believe") echoes the very distinction I draw at a theoretical level.

That thought seemed promising, but I wanted a more rigorous empiri-cal test. I got my chance a few years later when experimental psycholo-gist Larisa Heiphetz contacted me about our mutual interest in religious belief and its differences from factual belief. Together with my former research assistant Casey Landers, we designed some studies that would test whether Dan's observation was true—or at least true for American English. What follows is a compressed summary of our research.[5]

"Think" and "Believe" in the American Context

First, we used techniques from corpus linguistics. We explored the Corpus of Contemporary American English (COCA) to see whether its subcorpora (large bodies of naturally occurring text) revealed a greater number of *collocations* (significant associations) between "believe that" and overtly religious words than they did for "think that." There were striking differences: "believe that" had significant collocations with a range of religious and peripherally religious words (religious: *miracles, Allah, scriptures*; peripherally religious: *witches, celibacy*), and several ad-ditional religious words rose near the level of significance for collocations (*God, Jesus, baptism, sinful*).[6] By way of contrast, "think that" had no re-ligious collocations—and no religious words were even close to signifi-cance. That difference also held up in each of COCA's five subcategories considered separately (spoken, fiction, magazine, newspaper, and aca-demic subcorpora). We then approached the question the other way around: we held religious phrases fixed (e.g., *that God exists, that Allah exists, that God is, that the Bible says*) and checked whether "think" or "believe"[7] were collocates by searching immediately prior. The results were even more striking: fourteen of the twenty-eight religious phrases had "believe" as a preceding collocate; zero of twenty-eight had "think."

Second, we wanted to see if people would generate this word choice pattern spontaneously and under conditions that would rule out alternate explanations. We thus developed fill-in-the-blank sentence contexts for our behavioral second and third studies.[8] For example:

1. Dwayne _____ that Buddha found enlightenment while meditating under a bodhi tree.
2. Fred and Yuriana _____ that George Washington was the first US president.
3. Nick _____ that cassiterite is the chief source of tin.
4. Sharon _____ that she will meet her mother at the grocery store today.

For the second study, participants had a forced choice between grammatically correct forms of "think" and "believe." For the third study, they could freely enter any grammatically appropriate verb or phrase.

Before giving the results, let me point out a key design feature of the stimuli. The broad categories of sentence context were *religious* and *factual*, with our prediction being that participants would use "believe" more frequently in the religious contexts than in the matter-of-fact contexts.[9] But to address potentially competing explanations, we also had subcategories within the matter-of-fact contexts. We had attitude reports related to *well-known facts* (e.g., sentence 2), those related to *esoteric facts* not known to most people (e.g., sentence 3), and attitude reports related to *everyday life facts* (e.g., sentence 4). In each study, participants filled in blanks for a total of thirty randomized sentence contexts (fifteen religious, five well-known fact, five esoteric fact, and five everyday life fact attitude reports).[10] I will explain how this design feature addresses competing explanations shortly.

Results for the forced-choice sentence study showed a highly significant difference in the direction we predicted. Participants filled in 89 percent of the religious contexts with forms of "believe," but they filled in only 18 percent of the matter-of-fact contexts with "believe." So the following would be typical:

1. Dwayne <u>believes</u> that Buddha found enlightenment while meditating under a bodhi tree.
2. Fred and Yuriana <u>think</u> that George Washington was the first US president.

3. Nick <u>thinks</u> that cassiterite is the chief source of tin.

4. Sharon <u>thinks</u> that she will meet her mother at the grocery store today.

Furthermore, the difference between the religious contexts and the factual contexts taken together swamped the very small differences that cropped up between the matter-of-fact contexts.

To pick up the earlier thread, the differences within the category of matter-of-fact context allowed us to rule out two competing explanations for the main pattern in the data. One might speculate (a) that people were merely using "believe" in relation to contents they disagreed with; (b) that people were merely using "believe" in relation to contents they were unsure about. If (a) were true, we would have seen a lower incidence of "believe" for the well-known fact attitude reports than for the other kinds of matter-of-fact attitude reports since fewer people disagree with well-known facts. No such pattern emerged.[11] If (b) were true, then we would have seen a higher incidence of "believe" for the esoteric fact attitude reports since esoteric facts are more likely to seem uncertain. Again, no such pattern emerged.

Results for the free-response paradigm were just as impressive, if not more so. Even when participants could choose any word they wanted ("says," "knows," "thinks," "maintains," "hopes," etc.), they still used forms of "believe" for 51 percent of the religious sentence contexts. They used "believe" for only 8 percent of the matter-of-fact contexts.[12] And, again, the difference between religious contexts and matter-of-fact contexts overall swamped the small differences between matter-of-fact contexts.

To be clear, the point is *not* that the only use of "believe" is for religious credences—far from it! As I explained in Chapter 1, words like "believe" are flexible enough to apply to various mental state types and to be used for various purposes. The point, rather, is that there is a large and consistent pattern in the data, one that needs to be explained, and the explanation that fares best is our main hypothesis: that speakers are aware at some level of a difference in attitude type between religious credence and factual belief, and they use distinct attitude verbs in ways that track that difference.

We still need to consider one more possible alternate explanation for the data presented thus far. That would go as follows: the differential use of "think" and "believe" is to be explained entirely by the idea that people prefer "believe" for reporting attitudes with religious *contents* (or some closely related content type). If that were the whole story, no attitude difference (or sensitivity thereto) need be posited (at least for explaining these data).

Having anticipated that potential alternate explanation, we designed a final study with vignettes. Vignettes would allow us to hold the reported attitude contents constant while varying the surrounding contexts to make them religious or not.

Here are two of the vignettes we used:

> Sheila used to have a little shrine to Elvis Presley set up in her garage with candles and photos of Elvis. She even had a life-size statue of Elvis in flashy concert attire. On Elvis's birthday, she would take her guitar out and play his songs. During this time, Sheila _____ that Elvis was alive.

> Samantha began her mission to study penguins in the Antarctic in June 1977. She would spend most of her time in the Antarctic where she didn't even have access to the news. So she didn't see the headlines about Elvis Presley's death. During this time, Samantha _____ that Elvis was alive.

Each participant responded to a total of ten vignettes (five matched pairs in randomized order) by selecting between grammatically correct forms of "think" and "believe." As you might predict, our results showed that participants were more likely to put that Samantha (the Antarctic explorer) <u>thought</u> that Elvis was alive, and that Sheila (the one with the candles in her garage) <u>believed</u> that Elvis was alive. Since the attributed attitude contents were the same (*that Elvis was alive*), differences in reported mental state *content* couldn't be what explained the pattern of difference in verb choice.[13] The effect size here was not as large as in our sentence completion tasks, but it was still highly significant: participants selected forms of "believe" for 74 percent of the religious contexts and for 38 percent of the matter-of-fact contexts.[14] I think the best interpretation of these data is as follows: participants mostly viewed Antarctic explorer Samantha as having a (mistaken) factual belief that Elvis was alive; they viewed shrine-owning Sheila as having a religious credence with the same content; they were much more inclined to use the relevant form of "think" to report the factual belief; they were much more inclined to use "believe" to report the religious credence.

In sum, the phenomenon Dan observed in his classroom wasn't isolated; it appeared in multiple different ways in a large cross section of native speakers of American English. If we take the studies together, the only plausible hypothesis that explains *all* their data is the attitude hypothesis: that participants' differential word choice tracks and expresses different attitudes. Dirk, my Anglican minister friend, might well feel vindicated.

Two questions now emerge.

First, is this pattern idiosyncratic to speakers of American English, or does it appear in other languages and cultures around the world?

Second, given *that* people cognize a difference between religious credence and factual belief, what do they *do* with that difference? (What different expectations and norms do people have concerning the respective cognitive states?)[15]

I answer the first question in the rest of this section and the second question in the next.

Attitude Words around the World

Shortly after our data had been collected for the studies just discussed, I teamed up with a research group out of Stanford that sought to combine anthropological and psychological research on religious experience. The group's overarching hypothesis was that people in cultures with a "porous theory of mind" tend to have more striking spiritual experiences than elsewhere.[16] The field sites for this project were in the United States, Ghana, Thailand, China, and Vanuatu, which were chosen to enable comparisons using common religions in different cultures (mainly Christianity and Buddhism) as well as comparisons among different religions. Being part of this group gave me the opportunity to collaborate with Kara Weisman and Tanya Luhrmann on a crosscultural extension of the think-believe studies just discussed.[17]

In the United States, we focused again on "think" and "believe." For the other countries, we chose counterparts to "think" and "believe" by consulting native speakers and anthropologists with relevant expertise. In our Ghanaian field site, our field site managers and research assistants conducted the studies in Fante (an Akan language), with the focal words being *dwen* and *gye dzi*. In Thailand, our studies were in Thai and focused on คิด(*kit*) and เชื่อ (*chūa*). In China, our studies were in Mandarin (Standard Chinese dialect) and focused on 认为 (*rènwéi*) and 相信 (*xiāngxìn*). And in Vanuatu, our studies were in Bislama (an English-based creole) and focused on *ting* and *bilif.* Our hypothesis was not that these word pairs were exact semantic matches; rather, we hypothesized that people across cultures would use the pairs in parallel ways to express the same difference in attitude type (so, as before, our hypothesis was about social cognition—not about semantic content[18]). Essentially, we did a crosscultural replication of the earlier studies but left out the corpus linguistics component.[19] Our study stimuli, which were back-translated to ensure accuracy, reproduced the structure of the sentence context and vignette studies just discussed, though we made some adjustments to facilitate better understanding of the stimuli across the diverse cultures.[20]

Our first two studies were forced-choice and free-response sentence completion tasks, and the third was a vignettes study (as before).[21] Here are English versions of some of the sentence contexts we used:

1) Zane _____ that Jesus turned water into wine.
2) Greg and Theodore _____ that cell phones operate using batteries.
3) Astrid _____ that there are over 25,000 species of fish in the world.
4) Sharon _____ that she will meet her mother in the grocery store today.

And here are two of our vignettes:

Kerry had bad headaches in the afternoons all last year. Sometimes, her friends offered her aspirin. But Kerry took courses at a medical school. That school teaches that drinking water is the way to cure a headache and aspirin is not. So Kerry always refused the aspirin her friends offered. That's because she _____ that aspirin is not a cure.

Terry had bad headaches in the afternoons all last year. Sometimes, her friends offered her aspirin. But Terry belonged to the Church of Christ Scientist. That church teaches that prayer is the way to cure illness and medicine is not. So Terry always refused the aspirin her friends offered. That's because she _____ that aspirin is not a cure.

The logic of the third study, again, was to vary context while holding ascribed attitude contents constant (in this case: *that aspirin is not a cure [for headaches]*) to ensure that attitude rather than content was driving the effect we found.[22]

In fourteen out of fifteen experiments (three studies in each of five locations and languages), the data were significant in the predicted direction. That is, participants in all five different languages were more likely to use "think" (or its counterpart) for matter-of-fact attitude reports and more likely to use "believe" (or its counterpart) for religious attitude reports, and the difference between religious and matter-of-fact contexts swamped the minor differences between different types of matter-of-fact context. That participants responded so similarly in all locations, is the important pattern in need of explanation.[23]

The one experiment (out of fifteen) that didn't yield a significant effect was the vignettes experiment in Ghana. It's also worth noting that our effect size in Ghana for the forced-choice sentence study was somewhat smaller than in the other locations, though it was still significant. The difference could just be an experimental artifact, but another possibility is that in Ghana, having factual belief attitudes toward supernatural contents is a more common cognitive state since thought and talk about the supernatural are more common in Ghana than in our other field site locations.[24] That would explain our attenuated results at that site since it would explain why participants used the "think" counterpart (*dwen*) more often in connection with religious attitude ascriptions as compared with other sites. Clearly, however, more research on this particular topic is needed.[25]

Whatever the cause of the somewhat weaker pattern that surfaced in Ghana, it is still true that participants everywhere differentiated matter-of-fact and religious cognitive attitude ascriptions. And for everywhere other than Ghana, we were able to use the vignettes study to rule out the idea that the differential word choice was driven by content (or any content-related feature, like observability) rather than attitude. The best explanation of the broad, crosscultural pattern in the data remained our own.

As we put it in our final paragraph:

> Our results are most parsimoniously explained by our main hypothesis: Matter-of-fact belief and religious belief are distinct cognitive attitudes, and people in many different cultures and language communities are aware of the difference. The cognitive flexibility needed to utilize and

differentiate these attitudes is not specific to Westerners, Christians, scholars or some other rarefied group; instead, it appears to be widely shared.
Matter-of-fact beliefs are likely used in a problem-solving way to achieve
practical goals, while religious beliefs are used in guiding symbolic actions
expressive of sacred values (Atran & Axelrod, 2008); thus, tracking the
distinction between them may allow people to better understand and predict others' behaviors. Indeed, this distinction may be one of the common
features of a theory of mind that we increasingly understand to be subtle
and sophisticated across social worlds.[26]

3. WHAT ARE PEOPLE DOING WITH THE DIFFERENCE?

So far, we've seen that a great many people across diverse cultures are sensitive to a difference between religious credence and factual belief.[27] But
how do they regard the two attitudes differently, over and above choosing
different words to report them? People might have distinct *descriptive expectations*—different expectations concerning how people will behave if
one attitude is present versus another. People might also have different
normative views—different views about when having one attitude versus
the other is correct or good, or different views about how it is appropriate
to relate to factual beliefs or religious credences in themselves or in others.

Though there is bound to be substantial variation from person to person in this region of social cognition, both commonsense reflection and
psychological research reveal some patterns worth noting.

I've often heard people say things like *There's no point in arguing with
someone's beliefs!* Also: *It's disrespectful to argue with someone's beliefs.* People
who say the first seem to think that one is unlikely to change someone's
"beliefs" through disagreement, which is a descriptive expectation relating
to those "beliefs." People who say the second think it would be in some
way wrong or inappropriate to argue with someone's "beliefs," which is a
normative view.[28]

Both sayings have intuitive appeal. Yet the only way to make them
nonabsurd is to regard them as *not* referring to factual beliefs. After all, if
someone factually believes that *p* and another person possesses evidence
that *p* is false, it is often effective and appropriate for the person with
the better evidence to disagree and thereby improve the other's factual
beliefs.[29] This is true for banal factual beliefs (one person disagrees with

another's factual belief about how late the supermarket is open), as well as esoteric factual beliefs (one astronomer disagrees with another on her calculations of the onset time of the next lunar eclipse).[30] Updating one's factual beliefs in light of disagreement/correction from others is so normal and often quick that one frequently doesn't notice that that's what happened ("Ah, so it closes sooner than I thought!"). So, again, both of those sayings can only be charitably interpreted by assuming that they are *not* meant to apply to the attitude of factual belief: people saying them take the word "belief" as limited in roughly the way Dirk did—at least, in the context of the sayings themselves.

More formally, consider the following contrasting expectations:

Expectation 1 (change likely): if a person has cognitive attitude X toward proposition p, then disagreeing with that person about p (say, by presenting arguments or evidence to the effect that p is false) is at least *somewhat likely* to extinguish X.

Expectation 2 (change unlikely): if a person has cognitive attitude Y toward proposition p, then disagreeing with that person about p (say, by presenting arguments or evidence to the effect that p is false) is *highly unlikely* to extinguish Y.

Many contextual factors will affect the activation or deactivation of either one of these expectations, but it appears that almost everyone at least tacitly regards a large portion of ordinary factual beliefs as satisfying the X place in Expectation 1 (change likely). But a great many people regard religious credences as satisfying the Y place in Expectation 2 (change unlikely), which is what they mean to express by the first saying (*There's no point in arguing with someone's beliefs!*). Given what we've seen in preceding chapters about the *actual* difference in evidential vulnerability between factual belief and religious credence, it's unsurprising that many people should have such differing expectations: it just shows that they're paying attention.

In parallel fashion, the norms that many people subscribe to also seem to differ:

Norm 1 (appropriate): if a person has cognitive attitude X toward proposition p, then disagreeing with that person over p can be *appropriate*.

Norm 2 (inappropriate): if a person has cognitive attitude Y toward proposition p, then disagreeing with that person over p is *inappropriate* or *unseemly*.

My view is that most people regard Norm 1 (appropriate) as applying to factual beliefs, but many regard Norm 2 (inappropriate) as applying to religious credences (which is what gets expressed by the second saying: *It's disrespectful to argue with someone's beliefs.*). Furthermore, some awareness of Norm 2 (inappropriate) is often present even in people who would explicitly reject it: the strident atheist, say, who goes out of her way to counter religious credences with arguments and evidence typically knows that she is being provocative—indeed, that may be what makes it fun for her—which suggests she is at some level aware she is going against the grain of *other* people's norms, including Norm 2. Of course, there are many other norms that people subscribe to concerning how and when it is appropriate to disagree (disagreements should be potentially helpful, etc.), but that is consistent with my view that Norm 1 versus Norm 2 constitutes an important difference in how many people relate to factual beliefs versus religious credences.

Larisa Heiphetz and other colleagues of hers have done important experimental work that bears on just this topic.[31] They looked at people's attitudes toward *disagreements* involving three different kinds of belief: factual beliefs, ideological/religious beliefs, and preference beliefs.[32] They gave their participants scenarios where two parties express contradicting beliefs of one sort or the other, and then they asked their participants whether *both* parties could be right or *only one* could be right. For disagreements involving factual beliefs, they found that both child participants and adult participants held that only one of the disputants could be right. But for people disagreeing about a preference belief (e.g., *that blue is the prettiest color* versus *that green is the prettiest color*), participants mostly said *both* could be right. However, for religious belief disagreements (e.g., *that there is only one god* versus *that there are many gods*), participant responses were mixed. The average way of viewing such disagreements was *in between* factual belief disagreements and preference disagreements: many participant responses concerning religious belief "disagreements" indicated *both* disputants could be right, but many participants said *only one* could be; even within individual participants, responses were often heterogeneous, and, importantly, that pattern of mixed response appeared already among children. Heiphetz and colleagues interpreted this to mean that their participants viewed religious/ideological beliefs as being an "amalgam" of preference belief and factual belief, though they did not spell that view out further.[33] The important point, in any case, is that participants were applying different *norms* in relation to the differ-ent "belief" attitudes since judgments concerning whether someone can be *right* are normative judgments. Furthermore, it must be easy to pick up on this difference in norms since participants as young as five already showed significant differences in how they responded to religious versus factual beliefs.

My interpretation of these results is that sometimes, participants were responding to the *contents* of the religious disagreements (in which case—logically—only one set of contents in the disagreement could be true), but other times, participants were focusing on a different *attitude* type when it came to religious "beliefs," one on which it might be un-seemly to take sides when there's a disagreement. Participants who said *both* parties could be right may have been doing so as a way of upholding something like Norm 2 (inappropriate). Relatedly, if people saw attempts at resolving the disagreements as futile due to having Expectation 2 (change unlikely), saying both could be right may have reflected a default conciliatory stance for such situations. I grant that that interpretation is speculative. But the bigger picture is what's more important: the idea that people harbor different expectations and norms concerning factual beliefs and religious credences helps explain the interesting data that emerged in these studies.

A related cluster of different norms people have for "beliefs" already came out in the last chapter. Recall the finding of Emily Liquin, Emlen Metz, and Tania Lombrozo that many participants in their studies held there was a *need for explanation* when it came to scientific beliefs, but much less so when it came to religious beliefs. Relatedly, their participants also had a much higher level of mystery acceptance when it came to re-ligious beliefs than for scientific beliefs. In the last chapter, I took this as giving an indirect argument for my first-order view that factual beliefs are evidentially vulnerable, while religious credences are not.[34] But the same findings are even more relevant to the present discussion since they *directly* show that people subject religious credences and factual beliefs to different norms, along the following lines:

Norm 3 (need for explanation): if a person has cognitive attitude X toward proposition p, then that person (or someone from whom that person acquired that attitude) *should* be able to explain why p is the case.

Norm 4 (mystery acceptance): if a person has cognitive attitude Y toward proposition p, then that person's regarding p as a mystery is *acceptable*.

One way of explaining Liquin and colleagues' results is to say that many participants regarded factual beliefs (of which many scientific beliefs form a subset) as satisfying the X in Norm 3 (need for explanation) and religious credences as satisfying the Y in Norm 4 (mystery acceptance)—but vice versa only to a far lesser extent.

To summarize this section, there is reason to think that people's social cognition of factual belief and religious credence extends beyond mere sensitivity to the difference. Expectation 1, Norm 1, and Norm 3 are applied to factual belief much more than religious credence, and Expectation 2, Norm 2, and Norm 4 are applied to religious credence much more than factual belief.

Yet it is important not to look on this pattern of difference too rigidly. People often *do* quarrel with the religious credences of others, and in so doing, they may be simultaneously questioning the norms that people have about disagreeing. Conversely, there are many contexts in which one might find it inappropriate to disagree with a factual belief or in which (due to an explanatory chain's coming to an end) there is no need to explain the factual belief. This is a messy and contentious fragment of human social cognition. Furthermore, at the time of writing, we still await crosscultural extensions of the studies just discussed; many questions remain. Still, the important thing for present purposes is that this research reinforces the idea that people do cognize factual belief and religious credence differently, and they are often inclined to do something with that difference by having differing norms and expectations. Moreover, the differences we've just seen make sense in light of the first-order properties of the attitude types that we saw in previous chapters. Why, for example, disagree with an attitude that isn't evidentially vulnerable in the first place? And they will also make more sense in light of what we learn in the next two chapters, which concern the relations between "belief," identity, and sacred values.

4. THE WAY FORWARD

There are a few loose ends to tie up before we move to the next chapter.

First, let's return to Alison Gopnik's remark: *One way to tell that some-one doesn't believe something is if they say they "believe" it.* We can now make sense of this. The second "believe" in Gopnik's line tracks the use of "believe" in lay speakers, which typically does *not* express factual belief, as we've seen here. But the first (underlined) "believe" tracks philosophical use and thus (deliberately) designates factual belief (the straightforward attitude Fred has toward the existence of beer in his fridge). Gopnik's point is that one's saying "believe" tends to indicate *absence* of factual belief. Conversely, as Bernard Williams points out, one who factually be-lieves that *p* typically just says "*p*" rather than "I believe that *p*."[35] This is not a hard-and-fast rule, but it's clear by now that the *default* uses of "be-lieve" and "belief" are different for philosophers and laypeople.[36]

Second, one might wonder how this difference in patterns of use arose. Here, I recommend Wilfred Cantwell Smith's *Believing—An Historical Perspective.*[37] His second chapter gives a history that covers many twists and turns in the use of "belief" in English over the centuries. In short, he argues that "believe" in earlier stages of English meant *to hold dear,* which is different from the dry cognitive sense most philosophers give that term. If some of that earlier *hold dear* sense lingers in contemporary connotations of "believe," that makes it a natural choice for reporting religious credences and related ideological states—as opposed to straightforward factual belief.

Third, what is the methodological import of this linguistic disconnect between philosophers and laypeople when it comes to the words "believe" and "belief"? I want to be clear that there is nothing wrong with mem-bers of an academic discipline using words in a specialized way, and that includes philosophy. Normally, this just involves a trade-off: the special-ization of a word brings expressive power at the cost of making it harder to communicate the expressed ideas to the outside world. However, there is a serious risk when it comes to "belief." My impression is that most philosophers are not as linguistically sensitive as my colleague Dan. Hence, most, as far as I can tell, *do not realize* that their default usage of "believe" and "belief," which is heavily shaped by epistemology and philosophy of action, comes apart from lay usage.[38] I myself did not realize this clearly until that conversation back in 2012. But the contrast couldn't be starker. A family friend recently said to me in a living room conversation, "When I hear 'belief,' I think of *faith*." Philosophers are apt to think of Gettier cases.[39] The methodological risk here is that philosophers—not realizing the disconnect—will import *their* analytic understanding of "belief" to try to make sense of situations where a different mental state type altogether is present.

There are two lessons with which I'd like to close the chapter. They are already to some extent familiar, though we can now see them in a new light:

1. Philosophers commonly argue over whether a certain controversial mental state (a religious attitude, a delusion, a dream, etc.) is "really" a "belief." If it wasn't apparent already, it should be apparent now why this is so often a futile endeavor. The language of "believe" and "belief" is flexible enough—and sufficiently at cross-purposes in different portions of the population—that, *without substantial further clarification of what is meant*, it is opaque what has even been won if and when one manages to crown a given mental state "belief."[40]

2. It follows that it makes sense to introduce terms of art, as I have done, that help regiment the heterogeneous space of mental states that loosely and pretheoretically get called "beliefs" since calling something a "belief" only gives a very rough initial cut and can be misleading when read in certain ways. As Dan Sperber points out in the context of anthropology, where "belief" is one of his terms of interest: "The vagueness or arbitrariness of these terms has been repeatedly pointed out . . . if we want proper theoretical terms in anthropology, we should construct altogether new ones."[41] The same point extends beyond anthropology to the cognitive sciences more broadly. So, constructing new terms and using a theory to give them sense is what I've done thus far in this book.[42]

We've now seen in some detail what the cognitive processing differences are between factual belief and religious credence (these can be summed up by saying religious credence is a secondary cognitive attitude). We've also seen that people are sensitive to the difference and correspondingly have different norms and expectations concerning the respective mental states, though they are hardly unanimous on these matters. It's time to move to the next phase of inquiry, which involves asking the following question: What does this distinct attitude type of religious credence actually *do* for people?

My answer, in a word, is identity.

CHAPTER SIX

Identity and Groupish Belief

1. INTRODUCTION: PRACTICAL TRUTH VERSUS SOCIAL SIGNALING

In this chapter, I explore a simple idea and add psychological complexity as we go. The idea in its most distilled form is philosophical acid that will cut through much confusion about "belief" that plagues the philosophical literature. As that confusion dissolves, we can build up a more nuanced picture.

The simple idea is this: *if a belief guides <u>practical actions</u>, it works best if it is true, but if a "belief" defines a <u>group identity</u>, then it can still work or even work better if it is not true.* The terms "practical actions" and "group identity" will become clearer in due course.

That last clause ("*it can still work or even work better if it is not true*") sounds startling. But, as Anthony Appiah argues in *The Lies That Bind,* the "beliefs" and narratives that define group identities are mostly myths—largely false—that collapse under rational scrutiny.[1] Tales of national origins, theories of racial essence, talk of this or that being "in my blood," legends of great ancestors, and (yes) religious supernatural mythologies—poke any of these sacks of identity-defining ideas and a spate of falsehoods gushes out. The point I add to Appiah is that—for the purpose of defining an in-group—lack of truth is often a feature, not a bug.[2] That is not to say that *all* identity-constituting "beliefs" are false or incoherent, only that, given their role, they often are, and not by accident.

Here's a rough sketch of why this is so: if an idea is true and verifiable, then (in absence of pressure to the contrary) most people will be inclined to factually believe it, but in that case, it won't be a good identity "belief," since it won't distinguish those who "believe" from those who don't. Could we, for example, form a cult around the belief that cats like tuna? Such a cult wouldn't work, because everyone believes *that*, and distinguishing "believers" from "nonbelievers" is largely the purpose of identity-defining "beliefs." So, for group identities, it is more effective for the relevant narratives or other "beliefs" to contain falsehoods—or improbable or incoherent ideas. These "beliefs" will then be effective as internal badges that can be variably revealed through symbolic behavior.[3]

This is all in stark contrast to beliefs that guide ordinary practical actions, such as beliefs about the cost of gas. These beliefs generally depend on truth (at least approximate truth) to guide successful action (like budgeting for the month): someone with false beliefs about the cost of gas would wind up with downstream errors in her budget and spending. And such practically used beliefs are, in any case, easily updated and shared across groups. Table 6.1 illustrates the rough distinctions I have in mind.[4]

Religious credence plays the in-group defining role; factual belief plays the practical action-guiding role. This basic difference largely accounts for the functional differences we have observed so far between religious credences and factual beliefs. My argument is thus an inference to the best explanation: the features of religious credence that distinguish them

TABLE 6.1

	Mundane beliefs	Groupish beliefs
Role in action guidance?	Guide practical actions that achieve specific goals	Guide symbolic actions that indicate group allegiance
Dependence on truth?	Truth is needed for the guided actions to succeed in obtaining goals	Truth is not needed, and some falsity may help make the "beliefs" more distinctive

from factual belief, and for which there is independent evidence, are best explained by the fact that religious credences define group identities. There is a theoretical bonus here too. An important question is: What distinguishes religious credence from other secondary cognitive attitudes, like, say, fictional imagining? The role of defining a group identity answers that: the properties of religious credences that distinguish them from fictional imaginings are those that serve their role in marking group identities.

2. TWO EXPLANATORY ROLES

Examples give a rough pass at what I mean by *practical actions*: obtaining food when hungry, traveling to a desired location, acquiring and maintaining shelter, fixing your toaster oven, notifying someone to pick your kids up from school when you're not available, and so on. For such actions, the truth of one's relevant factual beliefs is crucial to success. Having true factual beliefs about where food is allows you to select movements that get you to it. If your belief about where the cookies are is false, you don't get the cookies. Having true factual beliefs about *when* your kids get out of school allows you to tell your next-door neighbor when to pick them up. With a false belief, you would tell your neighbor the wrong time. Even if a few local false beliefs don't undermine action success, there still needs to be a large background of true ones for most practical actions to succeed. Fixing your toaster oven at least requires your belief *that you have a toaster oven* to be true (otherwise, you would be fixing someone else's). All this is captured by the Davidsonian slogan mentioned in Chapter 2: beliefs cause and rationalize actions that satisfy one's desires *if the beliefs are true.*

We can now specify one *explanatory role* that is associated with the word "belief" by many speakers of English—especially philosophers and cognitive scientists.

> **Mundane Explanatory Role**: cognitive attitude X causes, guides, and explains actions that will likely achieve their goals if the relevant instances of X are true, but they will likely fail or be less successful if the relevant instances of X are not true.

Factual beliefs paradigmatically satisfy this Mundane Explanatory Role. The Mundane Role (or something close to it) is what philosophers of action—from Donald Davidson to Michael Bratman—have in mind when they posit what *they* call beliefs. Thus, theorizing about the particular features of belief in the analytic tradition is often an attempt to characterize more exactly the mental states that play this Mundane Role. So we read Fred Dretske writing things like "Believing something requires precisely the same skills involved in knowing."[5] And we read David Velleman advocating this thesis: "What distinguishes believing a proposition from imagining or supposing it is a more narrow and immediate aim— the aim of getting the truth-value of that particular proposition right."[6] Both passages emphasize that beliefs are truth-tracking in some way, and this makes sense on their usage since (1) practical action success depends on the truth of the guiding beliefs, and (2) humans do succeed in many practical actions.

Many psychologists also implicitly associate the Mundane Explanatory Role with the word "belief"—similar to philosophers. Consider the "false belief" task. In this task, which has numerous variations, an experimenter shows a child a doll who puts some candy in Box A (or something like that). The doll then goes away, at which point something (often a mischievous other agent) moves the candy from Box A to Box B, unbeknownst to the doll. Then the doll comes back, and the experimenter asks the observing child participant where the doll thinks the candy is (or where the doll will look for it). If the child says Box A (where the doll last left or saw the candy), then this child has successfully attributed a "false belief" to the doll. Thus, developmental psychologists in the false belief task tradition associate the word "belief" with what I call the Mundane Explanatory Role: a belief on this usage is a cognitive structure that guides an action that will fail (trying to find the candy won't succeed) if the belief is false.[7]

As we saw in the last chapter, most ordinary speakers of American English do *not* immediately associate the words "believe" and "belief" with the representation of mundane things, like where the candy is (though they can if prompted). So let's revisit this question: What notion *do* lay speakers more typically associate with the word "belief"?

This brings us to the second half of the simple idea with which we started, which can help further explain the think/believe contrast that surfaced in the last chapter. Often, the "beliefs" people talk about *explain what identity group they belong to.* Anabaptists are those who "believe" baptism is only valid when people confess their faith. Mormons "believe" Joseph Smith was a prophet. Hindus "believe" (among other things) that souls are immortal. It is not just that having such "beliefs" is a common occurrence among said groups. Rather, having such "beliefs" in part *makes* one a member of the group (though many other things are usually involved as well). And not having the "beliefs," though it does not always strictly entail that one is *not* a member of the group, does render one's membership questionable. This is true even for groups that do not doctrinally emphasize "belief" in the way that most forms of Christianity do[8]: having certain "beliefs" (whatever that turns out to be) is still *part of* what makes one a member of the relevant group.

Such "beliefs" illustrate what I call the Groupish Explanatory Role.[9]

> **Groupish Explanatory Role**: having cognitive attitude Y in part constitutes and thereby explains someone's belonging to a social group associated with Y (in the sense of sharing that group's identity).

Laypeople often have in mind something like this Groupish Role when hearing or using the words "believe" and "belief." This is not surprising, given what we learned about the history of "believe" from Wilfred Cantwell Smith: namely, "believe" originally meant *to hold dear*. Sharing a "belief" with a group is a way of holding the group dear.[10]

Now we come to the essential point: *aside from the fact that the <u>word</u> "belief" happens to be alternately associated with both Explanatory Roles (in different contexts), there is no good reason to think that the mental states that play the Mundane Role and those that play the Groupish Role are the same.* If we set the word "belief" aside, it becomes immediately clear that it is an open question to what extent the mental states that play the Mundane Role have overlapping characteristics with mental states that play the Groupish Role. It is not definitional, a priori, analytic, or in any way necessary that there should even be a large overlap. The impression otherwise is a philosophical illusion stemming from tacit reification of the word "belief." The overlap may be negligible, and it is up to theoretical and empirical investigation to determine whether the properties that *enable* mental states to play the one role are similar to, or different from, those that enable them to play the other. A "belief" in the doctrine of the Trinity, for example, has played the Groupish Role millions of times throughout history; it is highly questionable whether it has *ever* played the Mundane Role or what that would even mean. Beliefs about the locations of cookies play the Mundane Role millions of times every day; it is unlikely that they play the Groupish Role often at all. Why should we expect the functional characteristics that constitute the attitude taken in the first case to be remotely similar to the characteristics of the attitude taken in the other?

Let's recall the analogy from Chapter 1: Venus and Alpha Centauri are both commonly called "stars" in respectable (albeit pretheoretic) portions of English. But they are radically different phenomena, which have one appellation—"star"—in common. Analogously, the various things called "belief" may have radically different natures. Going forward, I'll use the phrase "Mundane beliefs" to refer to those mental states (whatever they are) that play the Mundane Explanatory Role, and "Groupish beliefs" for those mental states (whatever they are) that play the Groupish Explanatory Role. Given that formulation: Mundane beliefs and Groupish beliefs may be as different in kind as Venus and Alpha Centauri.

Many theorists, however, do not consider that possibility. Many philosophers frequently write phrases like "*the* concept of belief," which tacitly import the assumption that there is only one thing there. One of two distortions tends to follow. First, some theorists reason as if *all* mental states that fall under "the concept of belief" play the Mundane Role. This distortion, which is apparent in much of Davidson's writing, portrays the human mind as all too rational. Second, other theorists focus on Groupish beliefs and regard them as paradigmatic of "belief." In conversations with social psychologists, for example, I have heard the idea dismissed out of hand that "beliefs" are rational at all. The distortion here is the contrary of Davidson's: theorists who focus on Groupish beliefs tend to miss the overall rationality of the fabric of Mundane beliefs.

We can start to dissolve those two distortions by applying a little of the acid with which we started; that is, by noting a striking thing about Groupish beliefs: *Groupish beliefs do not need to be true for the actions they motivate to signal group solidarity.* So, unlike with Mundane beliefs, to whose function truth is essential, for Groupish beliefs, truth is often irrelevant or even counterproductive. Considering this point in detail will give us a clearer view of why some "beliefs" appear so irrational while other "beliefs" are more coherent and update in largely rational ways—and will help us make sense of other striking differences as well.

3. FALSITY AND SOLIDARITY

Let's use a toy example to clarify the conceptual points.

Suppose you and I belong to a group that worships the Green God of the Mountain, while our rivals worship the Purple God of the Valley. Thus, you and I both "believe" that there is a Green God on the mountain. This Groupish belief will guide a cluster of behaviors: shared sacrifices to the Green God, putting on green face paint, telling stories of how the Green God helped our ancestors in past wars, and so on. One of the main aims of all this behavior, which may be conscious or not, is to solidify our identity as members of the same clannish group.[11] Importantly, these behaviors accomplish *that* aim, even though the "beliefs" about the Green God are *false*: there is no such deity in the world outside our heads! So, unlike practical actions, whose guiding Mundane beliefs generally need to be true for their aims to succeed (for example, to get the cookies), *symbolic actions,* which are expressive of Groupish beliefs, do *not* need their underlying Groupish beliefs to be true to succeed at signaling and affirming a person's group identity.

Let's grant that an important aim in sacrificing to the Green God— and doing it in the very particular "right" way—is to solidify one's standing in the group and to encourage others to solidify theirs. This action works to do this because proper execution of the sacrifice shows (i) awareness of and interest in the group's customs and (ii) willingness to engage in costly signaling: it indicates to other group members that one is willing to undergo personal sacrifice out of allegiance to the group.[12]

To fill out the example, *we* "believe" the Green God has four arms, so there are four stages to the sacrifice; *we* "believe" the Green God prefers goat meat, so we sacrifice a goat to feed the Green God (even though we eat the meat ourselves); *we* "believe" the Green God is green, so we paint our faces green. Do these Groupish beliefs about the number of arms, the meat preferences of the deity, and the color of the God need to be *true* for the sacrifice to play its solidarity-inducing role? Otherwise put, does the success of this symbolic action rely on there actually being a green goat-loving deity with four arms? *No: Groupish beliefs are free to be false without loss of efficacy at fostering group solidarity through the symbolic actions they generate.* If nothing else, I hope that readers of this chapter will meditate on that fact at length when considering whether Groupish beliefs and Mundane beliefs, despite loosely sharing the appellation "belief," are the same sort of mental state.

We can now make the following move: we can explain why many Groupish beliefs have the characteristics of religious credences. Before doing that, however, it is worth specifying what I mean by *group identity* and how "beliefs" can play the Groupish role in helping to constitute one's group identity.

4. WHAT IS A GROUP IDENTITY?

The word "identity" has numerous uses, so let's focus on the one of interest: *group identity*. Group identity is obviously more than *strict numerical identity*—the relation every entity bears to itself. Even rocks and trees are identical to themselves. It is also more than *personal identity*—whatever it is that makes someone the same person across time. One can change group identities and still be the same person. Group identity is socially richer. But what is it? Appiah poses the question well: "Creed, country, color, class, culture . . . what on earth do they all have in common?"[13] Focusing on specific subtypes, we could ask the question like this: *Catholic, German, Black, middle class, Luo: what does everything on this list, which designates quite different properties but intuitively hangs together as a list of group identities, have in common?* It is more than that they single out sets of people. The phrase "people who like walnuts more than cashews" also picks out a set of people, but it isn't a group identity (at least not yet!). Metaphorically, it has no *identity glue*; the items on the lists just given do.

There are other senses of "identity" we can move past, even if they have features in common with group identity. There is *individual presentational identity*: Marie likes to present herself as *the die-hard Bob Dylan fanatic*, John tries to come off as *the guy who just loves drumming*, Sarah never misses the opportunity to reveal that she is *an expert on Brie and other French cheeses*. Those are individual presentational identities. Of course, one can love Bob Dylan (etc.) without making it an identity, so the presentational identity is something more than just loving Dylan's music.[14] To invoke T. S. Eliot: to maintain an individual presentational identity is to "prepare a face to meet the faces that you meet."[15]

Another identity notion is *relational identity*. Barbara *is a mom* (relation to children), Jeff *is a teacher* (relation to students), Terry *is a chess coach* (relation to other chess players). Again, one can be in these relations without having them as an identity. One might be a teacher just to pay the bills. So having a relational identity is being in the relation (or aspiring to it) *plus* some identity commitment—whatever that amounts to. A relational identity can also be an individual presentational identity or not. When it is, one signals the identity to others. For example, one might say, "As a mom, I think . . . " or "As a chess coach, I find that . . . ," and so on.[16]

Relational identities include *social role identities* as a proper subset, like *being a conductor* or *being the life of the party*. These are relational identities because they involve relations to other people, but they are not had in relation to *specific people*. One is the life of the party for whoever is at the next party. Social role identities form a bridge to our main interest— group identities—because they are simple examples of identities crafted out of a relation to *variable* people (who may come and go) rather than specific ones.[17]

Before we move to our target notion of identity, let's set one more aside. An *imposed identity* is one where others categorize you socially in a way that you reject or are ambivalent about. A good example is the imposed *convict* identity that Jean Valjean finds himself stuck with even after he is out of prison. He thought he was through with being a convict; the rest of society imposed it. Often, sexual orientation identity labels are imposed. One may have complex sexual desires but nevertheless find oneself shunted by society into one category or another—gay, straight, bi, queer, and so on—and often such identities are more impositions than items of allegiance. For present purposes, I set imposed identity aside, as interesting as it is, because my aim is to explain how Groupish beliefs figure into constituting the group identities one *accepts* rather than rejects.[18]

We can characterize group identities in one of two ways. We can start with *features of individuals* that make them members of given identity groups. Or we can look at *higher-level features of the groups* and work our way down from that sociological perspective toward individual characteristics. Presumably, either starting point will eventually lead to the other perspective, so starting one way or another is methodological predilection rather than opposition. Since many features of individual group identities are psychological, and since this book is a work of psychological theory, I start with individual psychological traits that are components of group identities. In other words, I characterize having, adopting, and maintaining a group identity as having, adopting, and maintaining a certain psychological state that relates one in characteristic ways to other group members who share the group identity. Of course, some nonpsychological features often get one started in having a certain group identity—who one's parents are, one's skin color—but *identifying* with the relevant group is a still psychological trait that needs characterization.

Here I lay out the broad functional features of the complex mental states that form the psychological part of an individual's group identity (such as Catholic, German, Black, middle class, Luo, etc.). Let me give just a few qualifiers before getting started. First, I'm not giving a fundamental analysis here; rather, I am highlighting features of the psychological state that will be useful for clarifying how Groupish beliefs feed into group identity, which will in turn help us understand why religious credences, as a subtype of Groupish belief, are the way they are. Second, these functional features come in degrees; one can have this or that feature to a greater or lesser degree. It follows that a group identity is not an all-or-nothing property but one that can be more or less on many dimensions. I'm highlighting what I take to be the most important dimensions. Third—relatedly—the psychological state I describe here is mature, fully formed group identity. One may build up to this over time, such that it is not entirely clear *when* an individual acquired the group identity in question.

For ease of exposition in stating the theory, let's stay with our toy example of the group that worships the Green God of the Mountain (then see endnotes for relevant empirical references). Let's say they call themselves "Greeners." Our question is this: What makes *identifying as a Greener* a different kind of self-categorization from, say, *noticing I like walnuts more than cashews*? Both single out sets of individuals, but the first has an identity glue that the second doesn't. What are the elements of that glue?

Here are seven features that I take to be constitutive of group identity mental states.[19]

1. *Group identities have a dual direction of fit.* When I self-categorize as a person who likes walnuts, this is simply a description of myself. I am not trying to make it the case that this description is true of me—it just seems to be accurate or something I discovered about myself. So that self-categorization has a *mind-to-world* direction of fit: it is a mental state that conforms to (or aims at representing) what the world of my gastronomic preferences is like. However, I might in an aspirational way view myself as an ultramarathon runner: I'm not there yet, but this idea represents something I am *trying to be.* So the mental state that represents me as an ultramarathon runner has a *world-to-mind* direction of fit: it motivates behaviors that will make the world more like what the mental state represents (it gets me out running, etc.). Importantly, the psychological states that make up group identities have *both* directions of fit. When I self-categorize as a Greener (worshipper of the Green God), this *both* describes what I am *and* describes what I aspire to be. That is, the mental state is not merely a matter of noticing I have a certain property; it is also a matter of motivating me to be a *better exemplar* of that property— where "better" is cashed out by the following functional features.[20]

2. *Group identities produce a disposition toward public acceptance of a label.* If I identify as a Greener, then the label itself matters. My group identity mental state disposes me to accept that label (and not opposed ones) in front of other people—especially other group members. Two people may worship the same or similar Gods, but if they label themselves differently—"I'm a Greener. *He* is a Guy of the Green."—then they don't have the same group identity and may even belong to opposed groups.[21]

3. *Group identities represent quasi-arbitrary criteria of inclusion.* This feature is a generalization of the last. It is mostly arbitrary whether we use the label "Greener" or "Guy of the Green," but that choice may make a decisive difference as to the group with which we identify. Similarly, there may be many other criteria of inclusion that are represented in my complex mental state of group identity, such as differences in how a particular ritual is done, abstruse points of doctrine, which side of the river I pray on, whether I kneel with the left or right knee, and so on. Some subset of these will be represented by the group identity in my mind, and they will single out who "counts" as being a member of *my* group. These criteria of inclusion are *quasi*-arbitrary because often there is a narrative that seems to justify the criteria: mythohistorical glosses on mostly arbitrary properties that are taken as determinative of one's identity. We might say that *we* use the term "Greener" because that's what

the great prophet of the Green God called herself, which may be true or not. And even if it is true, the fact that we adopt *that* as determinative of group membership is still mostly arbitrary. Having a certain lineage (often mythical), the acceptance of a certain sacred symbol, willingness to reproduce the exact peculiarities of certain tribal chants—all of these are quasi-arbitrary criteria of inclusion that are represented by one's group identity mental state.[22]

4. *Group identities produce a disposition toward social litmus testing.* Group identities dispose one to engage in litmus testing to see who is "really" a member of the group. Litmus tests are behaviors that cause other people to reveal whether they satisfy the quasi-arbitrary inclusion criteria relevant to the group. Do they *really* perform the Greener sacrifice like they are supposed to? Are they really descended from the prophet that first called herself a Greener? Or do they at least say they are? Do they viscerally object to the killing of frogs (to which every true Greener is supposed to viscerally object)? In general, litmus testing is the epistemic side of the metaphysical criteria of inclusion: satisfying the criteria *in fact* makes you a member of the group (at least in your mind); litmus testing helps you figure out who else does as well.[23]

5. *Group identities generate signaling behaviors.* People often don't wait for a litmus test to show other group members (and perhaps nongroup members) what group they are in. They wear clothing that indicates allegiance, they say certain things that would only make sense if one subscribed to the group's "beliefs," they engage in the rituals in the right way, they argue with people from other groups, they commit acts of violence, they say certain prayers, they let others know they are saying certain prayers, and so on. Many of these behaviors will have ostensible other purposes, like (supposedly) bringing rain or curing an illness. But as we saw in Chapters 3 and 4, people largely don't rely on instrumental-appearing religious actions to accomplish their apparent ends. So what are they for? One aim is signaling to members of the group that one shares their group identity.

6. *Group identities encourage the development of* habitus. *Habitus* is basically an interwoven cluster of behaviors (what one does before a meal); manners (how to address elders); mannerisms (like nodding one's head in a certain way); clothing choices (which shade of green); and expectations about how people should act in the world, where that cluster is distinctive of a certain culture or subculture. *Habitus* involves doing many entirely practical things in a certain way—*this* way to prepare the goat meat, not *that* one—and so it manages to infuse daily life in a more

thorough way than official rituals or rites. Having a group identity motivates one to act in ways that conform to the distinctive *habitus* of the group; this will be awkward at first, but it will eventually be swift and automatic. *Habitus* thus becomes an involuntary signal of group membership to both insiders and outsiders. Acquiring a *habitus* is thus partly a way of burning those bridges that would allow one to leave the group since it automates and makes it hard to override the signals one sends that identify one as a group member both to insiders and outsiders. In short, acquired *habitus* leaves one feeling at home in the group with which one identifies and alien in other groups.[24]

7. *Group identities include values.* The values in group identities are various and often inchoate. But the most important and pervasive feature of group identity valuing is that one values members of the group more highly than nonmembers. They are regarded as more worthy of help, their lives are more worth preserving, and one often (though not always) *feels* a sense of well-being just for being in their presence. And members of rival groups are often *disvalued.*[25] This is so, even though people often pay lip service to valuing all humans equally. Group identity valuing is a deeper, more emotion-laden state than consciously espoused or verbally articulated values typically are. Other group identity values specify *norms* for group members themselves: they ought to pass the litmus tests, to engage in the ritual signals, and to conform to the *habitus* of the group, and so on. Thus, the values of a group identity often include special sanctions for people who present themselves as group members but fail to live up to what is required. I don't get mad at non-Greeners for failing to sacrifice a goat to the Green God, but I get angry with Greeners who fail to sacrifice. And thus the values of a group identity give rise to boundary policing that keeps other group members in line. After all, if they are to be valued more highly than other humans, they had better do their part.

These are some of the salient psychological differences between the neutral self-categorization of being a person who likes walnuts and the group identity self-categorization of being a Greener. We can now say what makes an identity *group* different from a *mere set* of people: (i) members of a group all have the same group identity psychological state to a significant degree, (ii) they generally recognize each other as members of the same group as well (that is, as satisfying their quasi-arbitrary criteria of inclusion), and (iii) they apply their values and norms accordingly.

Many more things could be said here. The psychological state of having a group identity grows and changes over time. And the exact features of the state will vary from individual to individual, even within the same identity group. But we have already explored enough about the psychology of group identity to return to our main topic, which is how Groupish beliefs figure into the constitution of group identities. This will in turn pave the way for explaining why religious credences have the features we've noted for them in Chapters 3 and 4.

Here is how Groupish beliefs help constitute group identities.

a. *Representing a narrative about what unites the people who have the group identity.* Groupish beliefs playing this role will have contents like this: *we* all have a certain essence, share certain ancestors, have been blessed by the same deity, are headed for the same afterlife, are different from other groups in a certain way that makes us superior, are inheritors of a certain land, and so on. This role for Groupish beliefs serves to make the label feel like more than a *mere* label. Thus, although "Greener" may be a mostly arbitrary label, representing all Greeners as having descended from the same prophet will make the label feel less arbitrary.[26]

b. *Representing quasi-arbitrary criteria of inclusion.* This is in some ways an extension of the last role. These Groupish beliefs will have contents like these: a *true* Greener has properties *x, y,* and *z*; a true Greener also has had at least one face-to-face encounter with the Green God; a true Greener becomes emotionally outraged at the killing of a frog; true Greener men are sexually virile; true Greener women are sexually modest; and so on. In general, then, Groupish beliefs represent a series of generic propositions (or semipropositions[27]) about what makes one a member of the group.

c. *One criterion of inclusion may just be having certain Groupish beliefs.* That is, one of the Groupish beliefs that fall into category b. may have contents like this: *a true Greener <u>believes</u> the Green God will one day destroy all other gods.* Having that particular Groupish belief helps make you a member of the group: one who Groupishly "believes" this is more of a Greener than one who does not. Alternately, one who does not Groupishly "believe" that Greener men are sexually virile and Greener women are sexually modest may be less of a Greener, for *that* person does not have entirely the "right" set of

Groupish beliefs. There is thus often a self-referential aspect to Groupish beliefs taken as a class: they represent *themselves* as being part of what makes one a member of the group.

d. *Structuring litmus tests.* Groupish beliefs can also support the design of litmus tests to see who belongs. Any one Groupish belief—say, that true Greeners have had face-to-face encounters with the Green God—can help structure various litmus tests: How well does so-and-so tell the story of her encounter? How similar is so-and-so's story of her encounter to other encounter stories we've heard? Did so-and-so *really* become emotionally overwhelmed when she told the story of her encounter? Part of the power of having Groupish beliefs as internal markings of identity, in comparison with fixed external markings, is that any given Groupish belief can be probed or expressed in countless different ways, which gives rise to variable litmus tests and variable symbolic expressions of identity.

e. *Structuring behavioral signals of group membership.* One simple way to indicate that one has the Greener group identity is to express the contents of Groupish beliefs. One might say things like "Ah yes, the Green God always enjoys eating his roasted goat!" Or one might improvise: "Last sacrifice, I overcooked the goat meat for the Green God, and I don't think He was happy about it." The ostensible purpose of the last comment is to advise not to overcook the Green God's goat meat. The deeper purpose is to signal that *I* have the same Groupish beliefs as *you*. This role for Groupish beliefs, more than any other, accounts for the infinite creativity of symbolic action: since any underlying Groupish belief can be expressed in indefinitely many ways—sentences, gestures, paintings, dances—the symbolic action-structuring role of Groupish beliefs is a great part of what accounts for the explosive artistic quality of group identities.[28]

f. *Representing contents that seem to justify Groupish values.* This role overlaps with role a. If our group is to be valued more highly than others, it had better be special. Groupish beliefs that play this role often have contents like these: *our ancestors were particularly heroic, the Green God chose us for this land, those who worship the Purple God of the Valley took our land without cause,* and so on. Some of these descriptive contents may be partly based in fact, but many of them won't be. They will almost always be distorted and stylized in ways that support the values and norms represented in the group identity psychological state.

This brings us to the point where we can define symbolic action.

> **Symbolic action** is representational behavior that expresses Groupish beliefs.[29]

I leave "representational behavior" unanalyzed since the idea is clear enough: behavior that involves the production of representations, including representations that involve the body and its movements as constituents. There are many kinds of representational behaviors that are not necessarily symbolic actions in the relevant sense: most instances of gesture, sign, pretend play, painting, sculpting, certain forms of dance, and (of course) much spoken or written language. All of these behavior types involve representation. But much representation-producing action, such as writing down a recipe for chocolate cake, is not symbolic behavior in the sense we are after. If, however, representational action expresses a Groupish belief, then it is symbolic, for it has been made a symbol that can now be counted as sacred—to whatever degree and for whatever limited period—to the group.[30]

Now we come back to the crucial point. Aside from role b, which specifically applies to Groupish beliefs that *constitute* their subject matter (who's in the group), there is nothing about these roles that pressures Groupish beliefs to represent independent reality accurately. Unification narratives (role a) often have entirely fabricated elements, and this may help maintain the sharpness of the group's boundary. And any Groupish belief needn't be true for one simply *to have it,* so role c provides no pressure on Groupish beliefs to be true. Furthermore, false Groupish beliefs can still structure litmus tests (role d). The Groupish belief *that frogs have a sacred essence* may be what generates the litmus test of whether people become outraged over the killing of frogs. But that Groupish belief is false: there is no such thing as a frog's sacred essence.[31] Also, *any* Groupish belief can structure symbolic behavior in ways that signal identity. Painting the Green God holding a goat in each of his four hands is a symbolic action because it expresses the Groupish beliefs that the Green God has four arms and that he loves goat meat. It doesn't matter whether there really *is* such a thing as a Green God with four arms for this symbolic painting to succeed in its signaling. False Groupish beliefs work perfectly well for structuring behavioral signaling that does its job. Finally, role f puts positive pressure on Groupish beliefs to include distortions of reality that amplify the apparent worth of group members.

In sum, the collective roles that Groupish beliefs play in constituting group identity put essentially no pressure on them to be true or to track evidence—and some of those roles even put pressure on them to be, as it were, allergic to evidence and truth.

5. HOW RELIGIOUS CREDENCES PLAY
THE ROLES OF GROUPISH BELIEF

Suddenly, many things about religious credences snap into place. Why are they insensitive to evidence? Why, if people take religious credences so seriously, are they inferentially curtailed? Why are they compartmentalized? And why are they voluntary?

The answer to these questions is that the otherwise-puzzling features facilitate religious credences' playing the Groupish roles just identified. (By way of comparison, factual beliefs are *ill-suited* to playing the Groupish roles.)

Why Groupish Beliefs Are Not Evidentially Vulnerable

Groupish beliefs are internal symbols of loyalty to an in-group. However, they frequently have descriptive contents on which evidence can in principle bear. But there are at least three reasons why Groupish beliefs are not prone to revision in light of contrary evidence.

First, I would be a poor group member if my Groupish beliefs could be extinguished by something so trifling as contrary evidence. Loyalty shouldn't waiver, so internal symbols of loyalty shouldn't waiver. Yet any attitude that is vulnerable to evidence *will* waiver with incoming data. So the mental states that play Groupish roles a through f are not likely to be evidentially vulnerable. Group identity plays the role of gluing *us* together *come what may,* so our Groupish beliefs had better be invulnerable to the slings and arrows of evidence.

Second—as I've been stressing—while ordinary actions fail if the guiding Mundane beliefs are false, symbolic actions that express false Groupish beliefs are no more likely to fail than succeed. That point has important implications. Since having false Mundane beliefs about where the cookies are leads to not getting cookies, Mundane beliefs must be evidentially vulnerable, on pain of one's *continuing* to look in the wrong place. But sacrificing a goat to the Green God is a symbolic action whose success at signaling group allegiance isn't undermined by the perseverance of false "beliefs" in the face of the contrary evidence. So there is generally *little or no practical pressure* on Groupish beliefs to be evidentially vulnerable.[32]

Third—and here, we return to our philosophical acid—if Groupish beliefs have contents that stand athwart the evidence, then they are more likely to be distinctive of one particular group. Their expression will send a "stronger signal" in the sense of ruling out more possible sources. Thus, the roles of encoding a distinctive narrative and of encoding contents that justify distinctive values (roles a and f) put heavy pressure on Groupish beliefs *not* to be vulnerable to evidence.

It is thus no accident that Groupish beliefs tend not to respond to evidence. That is, the Groupishness of a mental state and its evidential invulnerability will typically travel together—exactly the opposite of which is true for Mundane belief.

Why Groupish Beliefs Lack Widespread Cognitive Governance

Two related considerations help explain why Groupish beliefs are likely to lack widespread cognitive governance.

First, widespread cognitive governance on the part of Groupish beliefs would undermine many practical actions. Here's why: if Groupish beliefs had widespread cognitive governance, their contents would pervasively infiltrate the contents of Mundane beliefs, so since Groupish beliefs often have false or distorted contents (for the reasons given), and since Mundane beliefs depend on truth to guide successful practical action, governance on the part of Groupish beliefs would undermine many practical actions.

To continue our running example, the Groupish belief *that the Green God will give me a good harvest if I sacrifice a goat* would threaten to undermine the practical steps I take in farming if it had widespread cognitive governance. I might end up reasoning as follows: *well, I sacrificed, so I needn't bother digging irrigation ditches*. But the lack of widespread governance—most importantly, the lack of governance over factual be-liefs—means that I don't reason this way, so the Mundane beliefs that lead me to farm properly stay intact, which is why I put in the work. We thus arrive at the following interesting theoretical proposition: inferen-tial curtailment serves the purpose of preventing Groupish beliefs from undermining truth-dependent, goal-oriented practical action; inferences from Groupish beliefs are likely to be curtailed when one's embracing of those inferences would thwart everyday practical success.[33]

Second, for symbolic action to be successfully produced, one has to use the real-world materials at one's disposal, which requires continual reality tracking. And continual reality tracking would be subverted if Groupish beliefs had widespread cognitive governance since (as we just saw) Groupish beliefs tend not to be evidentially vulnerable. This point is similar to the one Paul Harris makes about the use of props in imagi-native pretend play and artistic creation: the physical reality needs to be acknowledged even if one is pretending it is other than it is. Likewise, if I am the keeper of the sacred frogs for the Greeners, I will have to feed and care for them as I would ordinary animals so that they do not die when I bring them out for display on sacred occasions. The Groupish belief *that they have a sacred essence that makes them immortal,* if it had widespread governance, would potentially undermine my clearheaded view of what it takes to keep them alive. It is not just practical action that would be un-dermined if Groupish beliefs had widespread governance; much symbolic action would be as well.

Why Groupish Beliefs Are Compartmentalized

Much of the same reasoning explains why Groupish beliefs tend to be practical-setting-dependent.

First, if one lets Groupish beliefs guide action no matter the situation (that is, if they were practical-setting-*independent*), then one's practical actions would be ill-fated too much of the time since Groupish beliefs are often false, as discussed. For practical purposes, it makes sense to be able to turn off a body of "beliefs" that contains extensive falsehoods. This ex-plains why fanatics and extremists often come to unfortunate ends: their abilities to turn off their Groupish beliefs and thus to curtail their tenden-cies toward symbolic action are broken.

Let's regiment these terms—"fanatic" and "extremist"—which designate continuous properties. To the extent that someone acts as if they are in the religious practical setting *all the time,* that person is *fanatical.* Someone who prays and consults God seven times a day is *more* fanatical than someone who prays once a day, who in turn is more fanatical than someone who only prays twice a week. For fanatics, the religious practical setting has become *broader,* encompassing more of life. But even a great fanatic needn't be extremist in her behavior. One can pray or signal devotion constantly without engaging in any violent or risky actions. To the extent that someone engages in violent or risky actions, one is an extremist. Note that on this way of regimenting the terms, not all extremists are fanatics in the defined sense. One could be a fundamentalist Christian snake handler, for example, who only acts devoutly one day a week—in the service in which he handles poisonous snakes. This person's behavior is extreme but compartmentalized. Other examples of extremists who are not fanatics (in the defined sense) come from Scott Atran's *Talking to the Enemy,* which is an anthropological study of Muslim terrorists.[34] Atran's book portrays, among other things, members of terror cells who act contrary to their religious dictates much of the time—drinking, smoking, womanizing, and using and selling drugs. Yet their symbolic actions that express Groupish beliefs about jihad are extreme in the strongest sense: they are deadly.[35]

A fanatic who is not an extremist isn't in so much danger. This person, in effect, pays a continual practical tax for being a fanatic just in terms of time spent on symbolic action. But the extremist is in danger—and more so to the extent that they are also fanatical. One way of being an extremist is this: *performing symbolic actions and acting as if they were entirely sufficient for achieving their ostensible practical aims.* So while most religious people who pray for healing also go to the hospital, an extremist might avoid going to the hospital and treat going as if it were a violation of a trust in God.[36]

There is something of an irony in all this. A One-Map Theorist will have a flat-footed explanation of the extremist's refusal to go to the hospital: that person simply thinks God will cure him and that going to the hospital is not effective. And yet that "explanation" misses the Groupish dynamic of the extremist behavior. For the extremist to be aware that his refusal to go to the hospital is a strong signal of group loyalty, he must also be aware that his refusal constitutes some sort of *risk* or *cost*. That means that, at some level, he must also be aware that hospitals are places where people *are* healed and that prayer is a far more uncertain prospect. If he were not aware of this, avoiding the hospital would just seem like the sensible thing to do, and it would not seem to him like a strong signal of in-group commitment: it would merely be choosing the more effective option. Ironically, the risky symbolic action that on the surface appears to be guided by the "belief" *that prayer heals better than hospitals* is also tacitly guided by the awareness *that hospitals heal better than prayer*. In short, costly signals reveal that the signaler also in some sense believes the contrary of what they are expressing: it is *against the backdrop of mostly accurate Mundane belief* that expression of Groupish belief stands out.[37]

This all helps explain why *most* people are not entirely fanatical or extreme in their Groupish beliefs—that is, why their Groupish beliefs are practical-setting-dependent in the sense discussed in Chapters 2 through 4: practical setting dependence on the part of Groupish beliefs allows ordinary practical life to proceed safely.

Second, the main aim of expressing Groupish beliefs is (for the most part) not to be achieved when no one else is around: the aim of solidifying one's standing in the in-group. That is not to say one will *never* express Groupish beliefs when others aren't around. After all, the development of the right *habitus* may require practice. One might perform a certain symbolic action awkwardly in public if one has not practiced it in private, so there is still social pressure to have individual sessions of symbolic action. Furthermore, people often *tell* others of the symbolic actions they perform on their own—"*I pray all the time*"—and such claims, about which most people do not want to be totally dishonest (though exaggeration is common), typically solidify one's standing as a group member.[38] Additionally, one may simply *like* performing some symbolic actions. Humans are creatures who like representational play—from make-believe games to drawing—and symbolic action is a form of representational play, even if it is "serious" in the sense of being identity-constituting. Nevertheless, one does not play all the time, and one needn't act symbolically all the time to be a group member in good standing, so there will be long stretches of life in which people simply set their Groupish beliefs aside and act on

Mundane ones since these provide a better route to accomplishing practical ends.

There is thus a *lack* of pressure on Groupish beliefs to be active in guiding behavior across the board (since often group members are not watching), and there is positive pressure on them to *not* be active, since keeping them active can undermine the achievement of practical ends.

Third (ironically), relying on Groupish beliefs across the board would *undermine* one's ability to produce props for symbolic action. One's success in farming, for example, can be used as a prop in a symbolic action: "I prayed . . . and then I got a good harvest!" This utterance is a symbolic action, which uses farming success as a prop. But if one relied on prayer exclusively for the good crop, the crop would likely be poor, which would also make it a poor prop. Recall Jon Bialecki's example of the young woman who was short on money. The prayer group members donated to cover her bills, and she then attributed her financial betterment to God. Émile Durkheim would see this as evidence that the idea of a deity stands for the social group itself. Durkheim seems to be on the right track, though he often appears to have a commitment about *reference* that I do not agree with: Durkheim thinks the idea of the deity refers to the group, whereas I think its semantic value is the entity described in the relevant sacred myths and doctrines.[39] Yet we needn't get bogged down in issues of reference just yet. The key is that the donations Bialecki describes were expressive of the Groupish belief *that God would solve the young woman's financial problems.* But for the symbolic act to work, Group members had to do the ordinary humdrum things of going to the bank and withdrawing money or writing checks, which seems to presuppose the very contrary of the Groupish belief in question (why withdraw money for the young woman if God is already taking care of it?). So— again, ironically—the very Groupish belief that was being symbolically expressed had to turn *off* for a time for the prop to be constructed (the woman's financial betterment) that would be incorporated into the symbolic action. In short, symbolic actors need to deactivate their Groupish beliefs to successfully fashion the props that they implicate in expressing those very Groupish beliefs—just like stage actors who deactivate their imaginings of being a certain character in order to fashion props that will make them seem more like that character.

Why Groupish Beliefs Are Voluntary

At the deepest level, the world is independent of how you want it to be. The world was there before you, and the vast majority of things that occur in it are not under your influence. Of course, you can perform *actions* that will change some things, but even then, once you've made the change, the world is how it is. It thus makes sense that Mundane beliefs should also not be under direct voluntary control: their contribution to action success depends on representing the independent outside world accurately, so if the outside world is not under the direct control of your will, Mundane belief should not be either. And if we regard factual beliefs as paradigmatic Mundane beliefs, that theoretical expectation holds up.

Yet which group one identifies with *is* in many ways—though not every way—a matter of choice. We form or at least maintain alliances willingly. We can choose to be a card-carrying union member or not. We can leave the political party or stay. We can voluntarily join a new church. And though we appear to be stuck with many categorizations that others impose on us (imposed identity), we can choose to embrace them or not. Since group identity is largely a matter of choice, it makes sense for Groupish beliefs to be under voluntary control.[40]

The role of justifying Group values is also supported by the voluntariness of Groupish beliefs. Narratives that portray one group as more valuable than others generally contain extensive descriptive falsehoods, as noted; just consider the many Groupish claims about territory "rights" around the world that have been built on false or dubious historical narratives. Insofar as Groupish beliefs structure assertions that justify valuing one's group more highly than others (role f), they have to be able to be invented, which requires voluntariness.

Improvised Groupish beliefs can also support *individualized narratives* that help make one a member of the relevant group. The overarching narrative of one's group—where it came from, who its founders are, who its deity is, what the failings of outsiders are—is general; for most group members, it doesn't mention them as individuals. So the individual with the group identity must find a way to weave herself into the narrative, and this calls for the invention of *personal* Groupish beliefs—in addition to the *general* Groupish beliefs that she holds.[41]

Finally, at the deepest level, one must be able to adopt Groupish Beliefs to satisfy the relevant criterion of inclusion: the simple having of said Groupish beliefs. As we've seen, it is not evidence that compels one to have Groupish beliefs that portray the myths and doctrines of the group (role c), so one must be able to "decide to believe" them.

Inference to the Best Explanation: Religious Credences Are Groupish Beliefs

We can now remark on three converging facts that are best explained by the thesis that religious credences are a species of Groupish beliefs.

The first fact is that much academic research—over and above commonsense—has unearthed important connections between religious "beliefs" and group identity. Appiah has an entire chapter titled "Creed," which delineates how religious "beliefs," labels, and practices constitute identity groups, in which people have more trust with one another and less trust and often hostility toward outsiders.[42] Recall the example of Appiah's mother, who had a hard time "believing" but still wanted to be an "ordinary Anglican." Evidently, "believing" is part of what makes one a member of that group, as in most forms of Christianity. A related lesson comes from Ara Norenzayan's "Big Gods" hypothesis, according to which psychological representations of moralizing, monitoring deities are a product of cultural evolution that enabled widespread cooperation across large social groups.[43] Norenzayan holds that sharing "belief" in such Big Gods enables people to cooperate who have never met, since people of the same religion represent the same moralizing deity as monitoring their transactions. This, on Norenzayan's view, enables societies to grow much larger than the tribal units characteristic of human prehistory. The details of Norenzayan's hypothesis are disputed since a number of large-scale human societies emerged without representations of the sorts of deities it requires.[44] Nevertheless, one descriptive claim that coheres with the hypothesis is almost certainly true; namely, that people who share "belief" in the same deity cooperate more with one another than with people who do not share such "belief," and Norenzayan musters substantial empirical support for that claim.[45] In anthropology, Richard Sosis and Candace Alcorta show that religious kibbutzim in Israel typically last longer than secular ones, and a strong predictor of how long the community survives is how many burdensome rules they have.[46] This proved to be a seminal finding that initiated the subsequent literature on ritual displays as costly signaling, according to which much of the personal sacrifice that comes with being in a religious group has the function of signaling one's commitment—signaling that one will not defect when things get challenging. Another important finding in their paper is extremely revealing: the addition of burdensome rules in *secular* kibbutzim does not work nearly as well for ensuring community longevity. Sosis and Alcorta identify religious "belief" as the difference-making variable: it is when kibbutzim have burdensome rules *and* shared religious "beliefs" that they last longest as communities. All that, of course, is the tip of the iceberg of such findings.

The second fact, if I can call it that, is that the features of religious credences that we found evidenced in Chapters 3 and 4 parallel the features that Groupish beliefs can be expected to have. To recap the main points: evidential invulnerability enhances signaling since it differentiates Groupish beliefs more sharply from what anyone not in the group would endorse; practical setting dependence makes the Groupish believer more attuned to when and where to express the Groupish belief and ensures avoidance of pitfalls; lack of cognitive governance keeps untrue Groupish beliefs from infecting Mundane beliefs that must be largely true for practical action to succeed; voluntariness allows the creative generation of indefinitely many novel signals of group identity. Thus, many Groupish beliefs will likely be secondary cognitive attitudes. Playing the role of Groupish beliefs can *explain* why so many religious "beliefs" *are* secondary cognitive attitudes—in other words, religious credences. Otherwise put, given the roles we would independently expect Groupish beliefs to play, the idea that many religious "beliefs" are Groupish explains why they differ from factual beliefs in the ways they do.

The third fact is that religious action is often a display without useful practical consequences. Crossing oneself, dressing a certain way, singing songs, chanting, public sacrifice, stereotyped execution of a religious meal, stylized ritual cleansing, abstaining from sex, extremist violence that is politically counterproductive to professed aims—all of these actions and choices have only the most opaque instrumental value, if any. However, if we regard them as expressions of Groupish beliefs—as symbolic actions—they fall into place as being outward expressions of "beliefs" that support group identities. Thus, the fact that so many religious actions are symbolic is best explained by saying that the underlying mental states that guide and structure them—religious credences—are Groupish beliefs.

This brings us to another major thesis, which, though unsurprising, could not be articulated properly without the foregoing theoretical work:

> **Groupish Credence Thesis**: religious credences differ from factual beliefs in the ways they do in part because they are Groupish beliefs.

Being Groupish is unlikely to be the *only* pressure on religious credence to differ from factual belief. Religious credences may also have non-Groupish imaginative functions in the lives of individuals, like enabling them to have certain *personal* experiences that they might not have had otherwise, as theorists from William James to Tanya Luhrmann have emphasized. One may simply find many aspects of life more meaningful when one gives them an imagined supernatural gloss. But this is compatible with the perspective of this chapter: such supernatural glosses are not likely to come from evidentially constrained factual beliefs, so the imaginative role that religious credences play in "personal religion"[47] (to use James's phrase) most likely *also* pressures them to have properties that constitute them as secondary cognitive attitudes as opposed to factual beliefs. But that is a topic for a different time. Playing roles in constituting group identities is, for the reasons just given, a strong functional pressure on religious credences to lack the four characteristic properties of factual beliefs.

6. A THEORETICAL BONUS: DISTINGUISHING RELIGIOUS CREDENCE FROM MAKE-BELIEVE IMAGINING

Any account of religious "belief" should distinguish actually adopting a religion from merely pretending to adopt a religion. Throughout history and into the present, many people have participated in religious traditions merely as pretense with the aim of attaining other ends. Politicians feigning religious conversions, converso Jews pretending to be Catholics in seventeenth-century Portugal to escape the Inquisition, young atheists participating in religious ceremonies to please their parents, spies trying to blend into religious communities in foreign countries, and even clergy members who are secretly atheist[48]—all such characters have existed and continue to exist (though many of them eventually transform into "genuine believers"). Call this class of individuals "religious fakers." Let's then call people who profess their religion sincerely—whatever that amounts to—"sincerely faithful." Of course, each class is heterogeneous, but there is a strong intuitive difference between them, even if the lines are blurry. One question a theory of religious "belief" should answer, then, is this: Given that the observable religious behaviors of a religious faker and of a sincerely faithful person for a given religion may be largely indistinguishable, what is it about their internal psychological states that distinguishes them?

Here, the One-Map Theorist at first seems to have an advantage over a Two-Map Theorist like me. One-Map Theorists can say that the "sincerely faithful" simply think (factually believe) their religious doctrines and stories are true, while the religious fakers do not (they only pretend to think those things are true). This seems like a clean way to capture the distinction in question. Furthermore, such a One-Map Theorist can offer what at first sounds like a potent challenge to my view. The challenge goes like this: *Van Leeuwen maintains that religious credences are in essential respects like fictional imagining in how they differ from factual beliefs, so he makes all religious people—or at least those that have religious credences—into fakers.* In essence, the charge would be that my system of thought about religious credence lacks the expressive power to capture the distinction in question.[49]

My response is twofold.

First, it is actually the One-Map Theorist who lacks a certain expressive power; namely, the power to characterize a certain type of sincerely faithful person that we all know to be common: the sincerely faithful person who openly doubts their theological and other supernatural doctrines. Recall Luhrmann's informant: "I don't believe it, but I'm sticking with it."[50] This is a person devoted to the religion, despite being unconvinced or at least unsure of its supernatural claims. If anything, this person is an even *more* impressive member of the sincerely faithful because she persists in her devotion despite doubt. The existence of such sincerely faithful people shows that factual belief in the doctrines and stories is not *necessary* for being a sincerely faithful adherent of a religion. More generally, the One-Map Theorist will have a difficult time making sense of why there is so much doubt among the ranks of the sincerely faithful.

Second, I now have the resources on the table for both (i) character-izing the distinction between religious fakers and the sincerely faith-ful and (ii) describing what is going on with members of the sincerely faithful who openly disavow factual belief (such as Luhrmann's infor-mant). Let's start with (i). The sincerely faithful are those who have group identities that make them members of the religious group in question. Most importantly, their self-categorization as being a follower of *this* god, *this* ancestor, *this* spiritual practice has a dual direction of fit: not only will they represent themselves this way; they will also feel internal normative pressure to *make themselves* better exemplars of such adherence. Furthermore, they find such adherence to be a thing of value, such that those who adhere appear to be doing something right and proper. Correspondingly, their religious credences have what I call *perceived normative orientation*: it will seem to them that acting in ways that express religious credences moves them in a direction that is good and beneficial—and away from that which is bad and fearful. As Pascal Boyer and Pierre Liénard put it, "People just feel that they must perform a specific ritual, that it would be dangerous, unsafe, or improper not to do it."[51] The religious credences of members of the sincerely faithful, therefore, are associated with such feelings, and sym-bolic actions expressing such religious credences tilt the range of such feelings in a more positive direction. Charles Taylor, a Catholic, writes: "Somewhere, in some activity, or condition, lies a fullness, a richness; that is, in that place (activity or condition), life is fuller, richer, deeper, more worth while, more admirable, more what it should be."[52] Thus, perceived normative orientation means that religious credences, for the sincerely faithful, seem to provide a route to that place of "fullness"—at least when those credences are active in guiding behavior during sacred times and in sacred places. The same cannot be said for the pretense-guiding imaginings of the religious faker.

The religious faker, then, lacks a group identity corresponding to the religion they outwardly profess. Rather, they pretend to have that group identity for the sake of fooling people who genuinely have it. Though they may imagine the stories and doctrines of their feigned religion, they do not feel normative pressure to act in ways that express these imaginings *except insofar as they intend such actions to convince actual members of the sincerely faithful that they (the fakers) are also sincerely faithful.* Symbolic religious actions to the faker are merely instrumental: the atheist priest needs to keep a job, the spy needs to infiltrate, the atheist adolescent needs to avoid alienating her religious parents, and so on. In short, in the religious faker, the dual direction of fit and the higher valuation on being faithful (as opposed to not) are, in fact, lacking, though the faker pretends they are present.

Let's now turn to (ii): the sincerely faithful person who with awareness lacks factual belief that the stories and doctrines of their religion are true ("I don't believe it. But I'm sticking with it"). We already saw in Chapter 4 that, when probed, religious adherents even in religious countries, like the US and Iran, exhibit lower levels of confidence in the supernatural entities of their religions than in invisible scientific entities, such as electrons. Thus, it is likely that many sincerely faithful people lack factual belief in the main ideas of their religion without being metacognitively aware of that lack. Yet some, such as Luhrmann's informant, *are* aware that they lack factual beliefs with religious contents. So why are these people not religious fakers? The answer is that they possess the relevant group identity to a high degree, along with all the internal characteristics that go with it. Furthermore, their religious credences continue to play the identity-constituting roles of Groupish beliefs, regardless of their metacognitive awareness that those "beliefs" are not factual beliefs. (I suspect more people are like this than admit it.)

On this picture—to step back—there will not be a clear, bright line between those who are sincerely faithful and those who are religious fakers. Rather, one is sincerely faithful *to the extent that* one has the religious identity of the group and has religious "beliefs" that play their identity-constituting roles. There is a distinction to be drawn—and an important one—but it has an extensive gray area. The Anglican priests who do not think God exists probably do have some measure of Anglican identity, even though they spend much time pretending. They fall in the gray zone, though perhaps more to the faker side. The conceptual framework of the One-Map Theorist will have a hard time making sense of the gray area between faker and faithful since, for such a theorist, factual belief in the core stories and doctrines will be present or it won't. By way of contrast, the ability of my conceptual scheme to characterize this gray zone is a stroke in its favor. And a broad swath of "sincere belief" can now be characterized as religious credence that defines group identity. It is a mistake—one that entails *many* distortions of actual religious psychology—to assume that the phrase "sincere belief," when used in a religious context, refers to factual belief.[53]

CHAPTER SEVEN

Sacred Values

1. THE GOALS OF THIS CHAPTER

In Plato's *Euthyphro,* Socrates asks the religious figure Euthyphro to define "holy." "Holy," he responds, means "that which is beloved among the gods."[1] Socrates replies that this definition falls into a dilemma since we can understand it in two different ways. Either (i) the gods' love *makes* something holy, or (ii) the gods love certain things *because* they're independently holy. If (i) is how to understand Euthyphro's definition, holiness ends up being arbitrary because *whatever* arbitrary X the gods happened to love (let your imagination run wild) would then count as holy. But that's absurd. Surely, the gods have *reasons* for what they love. This pushes us to (ii): the gods love certain things *because* they're holy. Holiness, on this option, is something that inspires love among the gods. But if (ii) is correct, then Euthyphro's definition still doesn't work because holiness would be antecedent to the love it inspires, in which case, the gods' love, since it is merely a response to holiness that exists independently, couldn't be the definition of it.

In contemporary metaethics, this Euthyphro Dilemma is widely regarded as a checkmate against definitions of any sort of good (holiness, goodness, etc.) that appeal to divine approval. Does God's commanding

something make it good, or does God command it because it's independently good? If the former, then *goodness* would be the result of arbitrary
whim: God could have commanded whatever and thereby constituted
anything as good. But again, that's absurd: Could God have rendered
theft good just by commanding it? If it's the latter (goodness is the *reason* for God's commanding something), then God's commandment is
not the *definition* of being good, because the goodness of the things
commanded would be the antecedent and independent reason why God
commanded them. Otherwise put, either goodness is arbitrary (in which
case, why revere it?), or goodness is not defined by what God says or
thinks about it.

One might think this dilemma applies to *the sacred* as well: one cannot
define what it is to be "sacred" by reference to a deity's attitudes, because
a Euthyphro Dilemma would arise. My view, however, is that when it
comes to the sacred, Euthyphro had almost the right approach—with one
twist. The sacred is not that which is loved by the gods. Rather, the sacred
is that which is loved in a certain way by *us*; that is, what makes something sacred is that creatures like us hold sacralizing attitudes toward it.
And our representations of deities' attitudes toward that which *we* hold
sacred (along with our other representations of the supernatural) provide
us with an imaginative scheme for categorizing entities and events as having sacred status. In terms of the transparency metaphor developed in
Chapter 3, religious credences provide a cognitive map layer that lies atop
our factual beliefs like a transparency and dubs various actions, entities,
and events as the sacred ones.

And, yes, to leap on the first horn of the relevant Euthyphro Dilemma,
sacralizing attitudes are often arbitrary: any sacred object might not have
been if the sacralizing game had been played differently or if cultural evolution had wandered differently through the space of possible fetishes and
taboos, and many things that aren't sacred might have been and might
become so in the future.

What are sacralizing attitudes? So far, this book has answered the cognitive side of this question: religious credences are the identity-constituting
imaginative attitudes that generate symbolic actions and fold people,
places, and things into sacralizing doctrines and stories. Explaining *that*
has been my aim all along. But I would be remiss if I didn't also indicate

my approach to the conative side of the question since cognitions are idle without motivation. The way to phrase the other side of the question is this: What are sacred values?

With that setup, I have three aims in this chapter:

1. To familiarize readers with key findings in the psychological and anthropological literature on sacred values while assembling those findings in a way that amounts to a constitutive theory of sacred values as psychological states.
2. To show that my theory of religious credence meshes well with this empirical literature on sacred values; the plausibility of this mesh will lend further credibility to my theory of religious credence.
3. To use the idea of religious credence as I have developed it, to help *explain* some otherwise puzzling features of how sacred values manifest in human behavior.

To preview: just as religious credences differ from factual beliefs in fundamental ways, so, too, do sacred values differ from ordinary preferences in ways that leave much to be explained. The following hypothetical example, which illustrates some features of sacred values that have cropped up in the empirical literatures I discuss below, highlights some of those differences.

Suppose you and I live in the same apartment building and I stop by one day with an unusual request: I'd like to use your dictionary as a doorstop; it has the perfect weight and dimensions for this one door through which I need to move some furniture. You may be reluctant to lend the dictionary for such an odd request, but you would be more likely to say yes if I *added* some incentive, such as a bottle of wine. You might say, "For a bottle of wine . . . sure!"

Now suppose I make a similar request, but in this case, the book with the specific dimensions and weight is your King James Version of the Bible. I just need it as a doorstop. If you were like many devout people, you would be *much more reluctant* to let me use the KJV as a doorstop (as opposed to the dictionary) and would probably just refuse—and not because of the cost of the item. The KJV is, in a sense to be developed, sacred. So, in this case, you say *no,* citing that book's sacred status: "I can't let you use a sacred book as a doorstop." Now here's the key question:

TABLE 7.1

	Dictionary-as-doorstop request	Bible-as-doorstop request
Request with no incentive	Emotions: puzzlement, slight reluctance	Emotions: bafflement, offense, strong reluctance
	Preference: slight preference for not lending the dictionary	Preference: strong preference for not lending the Bible
Request with wine or other incentive added	Emotions: bemusement, happy anticipation of reward	Emotions: outrage, greater offense, possibly disgust
	Preference: *increased* preference for lending the book	Preference: *decreased* preference for lending, likely categorical refusal

How would you react if, on hearing your reply, I offered to throw in a bottle of wine?

If the KJV were sacred for you, you would likely be *more* outraged by the request coupled with the offer of wine than by the request itself. The extra incentive is an insult. You might say, "How dare you think I would desecrate my Bible for a bottle of wine?" If I then said, "How about two bottles?" you would probably become even more furious.[2] The two cases differ (i) in the emotions activated and (ii) in the effect that added incentive has on your overall preference scheme. Table 7.1 lays out the situation. Again, this example illustrates points that have emerged in the psychological and anthropological literatures on sacred values from the 1990s to the present.

First, humans (in most cases) have a way of valuing objects and outcomes that differs *in kind* from ordinary preferences. The outrage response to incentives is just one of the important differences. Let's call the cluster of psychological dispositions and capacities implicated in this way of valuing the *sacred values system*.[3] As a first pass to be filled in, we can say that the sacred values system treats certain things as inviolable (though in some ways, it is curiously [and seemingly paradoxically] quite flexible, as I discuss below). This contrasts with the *utilitarian values system*, which governs our decisions over things like how much money to spend on clothing, new curtains, or dictionaries. (I use "utilitarian" here in a general sense: roughly, having to do with that which is functionally useful. Think of the term as connoting "utility" in the economist's sense, not as referring to utilitarian ethical theory, though there are interesting connections between those senses.)

Second, the objects, entities, states of affairs, behaviors, places, ideas, persons, and events to which people's sacred values systems attach vary greatly from person to person and culture to culture. In the example case, you held your KJV sacred, but I did not. I likely find *other* things sacred. On a broader level, every culture has sacred values, but every culture has different sacred values. The sacralizing *way* of relating to things is (at least roughly) the same, but the things related to in this way are wildly different.

Third and consequently, sacred valuing is an attitude with no proprietary type of object or content: anything in principle can be sacralized or not, as I pointed out in Chapter 1. True, certain *kinds* of entities are more suitable for sacralization than others due to various psychological predispositions humans have.[4] But the massive variability in what people sacralize means it would be an error to define sacred values by reference to a *substantive* content or property. This is why the dilemma Socrates poses to Euthyphro is so tricky: one has the impulse to define that which is holy by reference to some substantive property, but what's holy is just a special subset of what's sacred, and since anything can be sacralized (or not), the impulse to point to some substantive property that is holiness is doomed to frustration.

Fourth, sacred valuing involves imbuing everyday, nonsacred objects with an *imagined* significance. We can suppose that the books in the two scenarios have the same dimensions, weights, approximate number of pages, and (probably) amount of dust sitting on them. But only one of the two rectangular assemblages of cardboard, paper, and ink has an imagined significance that is hard to spell out. Similarly, a patch of dirt is just a patch of dirt, but it is transformed into inviolable sacred ground in people's minds through imagined significance. Likewise for caves, stones, certain animals, or Styrofoam-textured bread wafers. When they are sacred, all these items have a dual aspect in the minds of those who sacralize them: there are their ordinary physical properties, about which one typically has factual beliefs, and there are imagined properties that in one's mind make them sacred. The relevant imaginings here are religious credences.

2. A CASE STUDY IN SACRED OBJECTS: THE GOLDEN PLATES OF THE BOOK OF MORMON

Before plunging into theoretical discussion, I'd like to examine an historical case to give more visibility to the notion of imagined significance. My treatment of it draws on Ann Taves's book *Revelatory Events,* though I gloss it using my own framework.[5]

According to the Mormon Church, Joseph Smith translated the Book of Mormon from ancient golden plates that the Angel Moroni had revealed to him. Both LDS and non-LDS historians, however, acknowledge that the physical setup of Smith's "translating" did not involve looking directly at any tablets on, say, a table in front of him. Rather, he would bury his face in a hat that had the "seer stone" in it. He would then "see" the golden plates *through* the seer stone. With this form of "vision" achieved, he would speak his "translation" aloud, and Martin Harris or Oliver Cowdery (inner circle members of the burgeoning church) wrote down what he said.

Were there any physical plates that people could pick up, touch, and look at without the special stone? There was, as far as historians can tell, a metal object of some sort (probably composed of metal plates), and this object, by Smith's orders, stayed hidden under a cloth. Smith's followers talked as if these metal plates were *the* golden plates, but, importantly, the cloth-hidden plates were not what Smith looked at while "translating," since his face was in the hat.

How, then, did members of the early church regard these actual physical plates, which were hidden by cloth and *not* used in translating? Taves writes:

> Joseph's wife Emma . . . [said] "I did not attempt to handle the plates nor, uncover them to look at them. I was satisfied that it was the work of God, and therefore did not feel it to be necessary to do so." . . .
>
> Joseph's directive [not to look at the plates under the cloth], understood by insiders as a divine injunction, functioned to set the plates apart in a Durkheimian sense. Although Joseph could supply the directive, others had to observe it in order for it to have any effect and, insofar as they did, they participated in the materialization of ancient golden plates. They did so by fusing an ordinary material object that could be viewed and "hefted"

and a non-ordinary believed-in object that could be seen only through the eyes of faith. In cognitive science terms, *believers linked their believed-in representation of the ancient golden plates with an ordinary, albeit concealed, material object, while skeptics did not.* Belief in the existence of ancient golden plates and special objects that gave Smith the power to translate them distinguished insiders from outsiders and thus played a crucial role in constituting the emergent group as a group.[6]

Here, we see a two-map cognitive structure in the minds of the early Mormon adherents. They were aware of a mundane metal object—the physical plates under the cloth—that could be cognized using the senses and about which one would have had factual beliefs (concerning its size, whereabouts, etc.). But they also represented the sacred, "believed-in" golden plates that had to be imagined, "seen through the eyes of faith," or looked at through a special seer stone in a hat. That Smith did not look at the object under the cloth when "translating" is telling: the entities that had to be viewed and translated were in some alternate realm—not just sitting on a table, even though the object on the table was revered *as* those ancient plates in the alternate realm. It is also telling in a different way: even in the sacred context of "translating," neither Smith nor his followers got confused about what the physical properties of the plates under the cloth actually were; in other words, their religious credence layer never had cognitive governance over their factual belief layer.

This cloth-hidden object, then, was a *prop* in Kendall Walton's sense.[7] Relative to a certain game and against a background of more general religious credences, that prop prescribed further religious credences, which would then guide symbolic behaviors in relation to the prop/sacred object. Carefully avoiding the removal of the covering cloth, for example, is one such symbolic behavior: the behavior represents the idea that *this object* has a divine injunction hanging over it. But there is also maintenance of the game happening in that very same behavior: one can surmise that part of the reason church members were so eager to comply with the directive *not* to look at the object under the cloth was that they were aware, though their group identity compelled them not to admit it, that it was *not* really a set of ancient golden plates at all, and they did not want to ruin the sacred play by seeing what was actually there.

Thus, religious credences relating to a set of humdrum metal plates in Upstate New York in the late 1820s both (1) categorized them as sacred by locating them in a supernaturalistic narrative and (2) represented contents that could be expressed through symbolic behaviors performed in relation to them, for example, carefully observing the taboo on lifting the cloth.

It is hard to overstate the arbitrariness of most of this. *Any* humdrum metal plates of the right dimensions *could* have played the appropriate prop role. There would have been nothing inherently special or sacred about any of them. They all would have had the utilitarian value of some regular metal plates. But *this* set of plates—the one under the cloth, not other ones—ended up having sacred value and prompted reverential behaviors accordingly. To relate this to the Euthyphro discussion above, the humdrum metal plates had no antecedent property that made them sacred; rather, it was the sacralizing regard—both cognitive and conative—that *made* them sacred objects. Importantly—and to connect this to the last chapter—Taves writes that this sacralizing regard "distinguished insiders from outsiders."

3. FEATURES OF THE SACRED VALUES SYSTEM

Some readers might have had the following reaction to the opening section of this chapter: *Look, it's well known that human preferences depart from the idealized utility functions of rational choice theory. Thinkers from Allais to Ainslie have shown incoherence or distortions in people's actual utility functions. What's the big deal with sacred values? Aren't they just more distortions in the mix?*

The answer is *no*. The departures from economically rational utility that behavioral economists have detected are mostly glitches among human preferences that *approximate* rational utility. Daniel Kahneman and Amos Tversky's prospect theory, for example, which formalizes many of the ways humans depart from economic rationality, is a *modification* of rational choice theory.[8] Those departures are best seen as bugs in a human utilitarian value system whose competence (if not always performance) aims at maximizing utility. By way of contrast, sacred values have the *job* of departing from economic utility—of motivating certain actions *come what may*. Its departures are features, not bugs.

Here, I identify six features of sacred values. In developing these notions, I use the term "entity" broadly to refer to objects, actions, agents, behaviors, traditions, places, events, outcomes—basically, anything that can be sacralized. Also, these features come in degrees and are logically independent of one another, so it won't be surprising to find cases where some are present but not others. Still, the six features cohere well enough to cluster together and form an attractor position in a psychological space that has markedly distinct qualities from utilitarian values.

First, there is *constitutive incommensurability.* Previously developed in a philosophical context by Joseph Raz, Philip Tetlock and his colleagues present it as an empirical finding about the psychology of sacred values.[9] They give the metaphor that humans have both an intuitive economist and an intuitive moralist-theologian living in their psyches. The intuitive economist helps one decide between new sneakers, new bedsheets, or a dinner at a decent restaurant.[10] The intuitive moralist-theologian tells one which things may not be traded in economic exchanges of any sort. Incommensurability implies that the sacred things stand outside the intuitive economist's utility function; there is no common internal metric for comparing the sacred value of, say, holy ground to the utilitarian value of any number of sheets, sneakers, dinners, or dollar amounts. *Constitutive* incommensurability implies even more: people find it outrageous *even to contemplate* trading something sacred (like holy ground) for something merely utilitarian (like money). The outrage response is the key.[11] One's intuitive moralist-theologian doesn't haggle with one's intuitive economist; he says *get out of my house.*

Constitutive incommensurability is a phenomenon that will cause decision theorists, economists, and economically minded psychologists to stub their toes. When describing "rational preferences," such theorists appeal to various axioms and properties.

One preference axiom is Completeness: *for all A and B, the individual prefers A to B, prefers B to A, or is indifferent between A and B.*

This means that a rational individual ("rational" on this framework) will be able to choose between any two outcomes—or just be indifferent between them. No one, of course, has considered every trade-off between pairs of utilitarian goods, but extensions of one's prior preferences come easily (Are these shoes really as valuable as a year's Netflix subscription?). But constitutive incommensurability implies that entities valued as sacred stand outside such economic comparisons in the mind of the person who has the relevant sacred values: *principles like the Completeness Axiom just do not apply.* A determined decision theorist might try to find a way of modeling this phenomenon: for example, by saying sacred outcomes are preferred over all else. But that move misses important psychological dynamics. First, taboo trade-offs (those that pit utilitarian values against sacred values) are *not* always resolved in favor of sacred values (as we'll see in the next section). Second, taboo trade-offs trigger outrage and hostility toward the person offering the trade: a different *kind* of response from that elicited by mere economic comparison. This is why one can't convince ultraorthodox Israeli settlers to move out of Palestine by offering to pay them: valuing the land as sacred forbids contemplating such a trade and may make one eager to *punish* community members who do contemplate it.[12] The reaction is far more aggressive than just saying, "Sorry, not enough money on the table."

In sum:

> **Constitutive incommensurability**: if a person's sacred values system attaches[13] to an entity, proposed trade-offs between that entity and merely utilitarian entities trigger outrage or disgust.

The second feature of sacred values also contrasts with economically rational preferences. In particular, economically rational preferences are typically *additive*[14]:

$$u(x, y) = v(x) + v(y)$$

This says that the utility of having both *x* and *y* is equal to the sum of the values of *x* and *y* considered independently (the utility of having sneakers *and* sheets is the sum of the value of having each separately). More generally, any additional positive outcome of an act should make performing it *more* desirable.

The dictionary doorstop scenario displays additivity. Say there is some disutility to letting me use the dictionary as a doorstop (call it -1) and some utility in having the bottle of wine (call it 5). Putting them together gives a utility of 4, so you lend the dictionary. But that's not what happens in the scenario where I ask for your KJV. The additional incentive there makes the request *more* outrageous and fulfilling it *less* desirable—the incentive is an insult.

Let's call this feature *incentive outrage.*

Incentive outrage: if a person's sacred values system attaches to an entity, additional utilitarian incentives to accept a taboo trade-off involving that entity trigger greater levels of outrage or disgust than that taboo trade-off would without the additional incentive, thereby making the incentivized version *less* acceptable to that person.

Jeremy Ginges and his colleagues show that incentive outrage arises among both Israelis and Palestinians for taboo trade-offs involving contested territory.[15] When study participants considered deals that would trade land for peace, a subset of individuals in each group ("moral-absolutists who had transformed the issues under dispute into sacred values"[16]) responded more harshly to peace deals that had "added instrumental incentive" (such as money) than to the peace deals by themselves: the "Taboo+" deals elicited greater anger and disgust as well as greater support for violence against the opposing outgroup. So incentive outrage is an opposition to additivity *packaged with* hostile emotions and behavioral tendencies (anger, disgust, aggression, etc.). It is a context-specific moral allergy: in sacred contexts, deal-sweeteners trigger revulsion.[17]

Two other features of economic rationality that sacred values contravene are *sensitivity to the probability of success* and *temporal discounting.* The first means that one becomes less likely to perform a given action if one learns it is less likely to achieve its goal; the second means that future rewards are valued less than present rewards. By way of contrast, when sacred acts are called for, their imperative is insensitive to the probability of success: if martyrdom is called for, it will still be called for if the devoted actor learns it is unlikely to accomplish its stated aim.[18] Relatedly, lack of temporal discounting means that a sacred outcome's being further in the future does *not* diminish one's commitment to performing acts in service of it: getting back the holy ground may be indefinitely far off in the future, but that in itself makes the devoted actor no less devoted.

This gives us:

Insensitivity to the probability of success: if a person's sacred values system attaches to an outcome, that person's imperative to perform certain actions in service of it does not diminish on learning that those actions are less likely to succeed.

No temporal discounting: if a person's sacred values system attaches to an outcome, the value that person places on it is insensitive to how far it is in the future.

These four features (constitutive incommensurability, incentive outrage, insensitivity to probability of success, and no temporal discounting) give substance to the notion of inviolability: people treat an entity as inviolable to the extent that their valuing of it exhibits these four features.

This brings us to the fifth feature. Part of the reason *why* sacred values imply inviolability is that they activate contagion thinking. Generally, when people intuitively categorize certain things as contaminants (e.g., cockroaches), they track them in ways that are insensitive to rational reflection (e.g., if X seems contaminated and X touches Y, then Y <u>seems</u> or <u>feels</u> contaminated too [even if one knows intellectually that both X and Y have been sterilized]).[19] Sacred values thus prompt revulsion toward anything that appears to contaminate sacralized entities. Contaminants can be almost anything profane, which is a big part of why ritual cleansing before entering sacred spaces is a common feature of religions. Describing his sacred values protection model (SVPM), Tetlock writes:

> Resource constraints can bring people into disturbingly close psychological contact with temptations to compromise sacred values. The SVPM predicts that decision makers will feel tainted by merely contemplating scenarios that breach the psychological wall between secular and sacred and engage in symbolic acts of moral cleansing that reaffirm their solidarity with the moral community.[20]

Contagion thinking for sacred values works in multiple directions. On the positive side, something touched by something sacred often becomes sacred itself (as in the myth of lotus flowers blooming in places where the Buddha walked as an infant). On the negative side, anything that appears to contaminate the sacred (such as a peace deal that would exchange holy ground for money) is shunned.

Contagion thinking: if a person's sacred value system attaches to an entity, nonsacred entities that come into contact with it either (i) contaminate the entity (which calls for cleansing of some sort) or (ii) are themselves made sacred by it (which then makes *it* an object of sacred value).[21]

Sixth and finally, recall from the last chapter that *group identities* include values as part of their psychological makeup. Relatedly, a commonplace in the literature under discussion is that sacred values play a Durkheimian role in constituting "moral community" and distinguishing "insiders" from "outsiders." Putting those points together, sacred values—those with the features just described—comprise many of the values in one's group identity: members of a group don't just happen to have the same or similar sacred values; rather, they're not proper group members if they *don't* have those values. Having the sacred values is a criterion of inclusion. Symbolic actions that express them signal group identity, and litmus tests probe for the sacred values in others (e.g., Is this person willing to *punish* someone else who violated a sacred value? If not, perhaps they're not *really* among the faithful[22]).

Furthermore, the fact that sacred values partly make up group identity helps explain their other features. In particular, the inviolability features of sacred values enable groups to function *as a unit*: the inviolability of sacred values fuses one with the group with which she identifies in a way that is robust against the many vagaries of convenience and individual advantage.[23] Thus:

> **Group identity constitution**: if a person's sacred values system attaches to an entity, treating that entity as inviolable is a criterion of inclusion in one's relevant group identity.

In sum, six features distinguish sacred values from ordinary utilitarian values:

1. Constitutive incommensurability
2. Incentive outrage
3. Insensitivity to probability of success
4. No temporal discounting
5. Contagion thinking
6. Group identity constitution[24]

The task in the next sections will be to spell out more exactly how religious credences, as cognitive states, interact with sacred values. One more point, however, is in order in this section.

An object's being valued by one system doesn't preclude its being valued by the other. On the contrary: sacred entities continue to be valued by the utilitarian values system under their mundane description. Consider our running example of the KJV. Even though a person's sacred values system attaches to that object, their utilitarian values system attaches to it as well. As Daniel Kahneman emphasizes in various places: people's preferences are not about things directly; their preferences are among things as *described in certain ways.*[25] Hence, the value a person places on an object varies with the frames under which the object is described. An E. E. Evans-Pritchard quotation we saw in Chapter 3 is illustrative here as well: "Some peoples put stones in the forks of trees to delay the setting of the sun; but the stone so used is casually picked up, and has only a mystical significance in, and for the purpose and duration of, the rite."[26] The sacred stone is just a stone outside the religious setting. In fact, the oscillation in valuing type that applies to (sacred) objects is an important and puzzling phenomenon that the present theory can help explain.

4. TWO PUZZLES: FAILING TO ADVANCE THE SUPPOSEDLY SACRED

Let's develop the concern just noted. If you read the sacred values literature in too cursory a fashion, you might come away with a false impression. You might think that anyone who values an entity as sacred will invariably do her utmost to honor and revere that entity at all costs, to preserve it, protect it, and promote it into the future. All opportunities will be taken in service of the sacred entity, and one will constantly seek ways to be guided by one's sacred values. Indeed, anyone who *has* a sacred value will seek to give that impression.

But that's not what we see if we look below the surface. Tetlock points out that "vexing questions remain" about the divergence between people's presentation of sacred values as being inviolable and "the compromises that they make in the real world of scarce resources."[27] Consider the two puzzles below as among those vexing questions.

People often *forgo* opportunities to further what they claim as a sacred value *if* the opportunities in question don't have the right form. In other words, instrumentally rational actions that *would* further certain supposedly sacred goals are often ignored or even shunned. A clear example of this comes from the American Evangelical prolife movement. The entities valued as sacred in this case are embryos and fetuses. And many dramatic actions are taken to protect those sacred entities: constant legal battles, protests outside abortion clinics, promotion of restrictive laws, stacking of courts, and more. We have seen the results of this work clearly in the fall of *Roe v. Wade*. But an obvious instrumentally rational approach to preventing the destruction of fetuses and embryos would be promoting social services for mothers, especially would-be single mothers: affordable or free daycare, paid time off from work, subsidies for continuing education, and so on. Such services would dramatically reduce the number of abortions since they would eliminate the hard choice many single women face between having a child and continuing with their career or education. Furthermore—in addition to the point's being obvious—empirical data support this line of reasoning. Countries like Germany, Belgium, and France have significantly lower rates of abortion than the United States, despite having easier access and more tolerant attitudes toward it.[28] Still, the majority of American pro-lifers neglect to advocate certain relevant policies that would further their stated sacred aims. And that is not the only example of this phenomenon. For decades, both Palestinian and Israeli leaders have neglected opportunities to make symbolic compromises that would instrumentally advance the aims that are supposed to be sacred to them (land in one case, security in the other). Examples can be multiplied. So our first puzzle when it comes to sacred values is this: *If sacred values are so important, why do devoted actors often shun clear instrumental means toward their stated sacred aims?*

Another puzzle is the fact that entities once deemed sacred can suddenly lose that status. The stones Evans-Pritchard describes are just one example of this. Relics and fetishes are commonly discarded or replaced even when the people doing the discarding are still devoted to the religion for which they stand. We know this must happen, but it is unfortunately not well documented. Pascal Boyer writes of the anthropological record on the phenomenon of discarding sacred objects as follows:

> That is one of those things that are strangely not well documented. I had noticed that with special decorated boxes that people use in rituals, but then let the children play with in other contexts. I think the Zuni have the

same with some of their "dolls." I mentioned that to several anthropologists, who said that they too had seen similar behavior. But no-one bothered to write about that. . . . Obviously more organized religions would provide more examples, e.g., recycled statues.[29]

One such "more organized" religion is Catholicism, whose practitioners have such a tendency to amass sacred objects (which must end up somewhere) that injunctions exist to the effect that discarding them must be done in a "respectful" way (burial, burning).[30] And often, a prescribed ritual object will simply be replaced by a rough substitute if the originally intended item cannot be found. This gives us our second puzzle: *If sacred values are so important, and if sacred entities are inviolable, why do many of the entities to which the sacred values system attaches so easily get discarded or replaced?*

5. HOW RELIGIOUS CREDENCES RELATE TO SACRED VALUES

The main theoretical burden of this chapter is to explain how people's religious credences relate to their sacred values. Furthermore, I should do this in a way that helps solve the puzzles just noted.

Here are the claims that bear these burdens:

> **Claim 1**: *many* religious credences people have describe supernatural entities that those people can't locate in space.[31]

Such supernatural entities and events include deities, spirits, saints, afterlife realms, miracles someone heard about from someone who knew someone who saw them in an exotic country on a mission trip, and so on. Let's call such religious credences *detached* because they don't have constituents that designate specific objects, events, or properties in the sensible world around the person who has them, nor do they designate anything that she could reliably find by moving about.[32] The narratives and doctrines given by detached religious credences are, in other words, freefloating relative to places and things that can be perceived: their supernatural subjects are in another realm, in the past, or in an uncertain location.

If *all* religious credences were detached, most people—even the most devout—would never feel like they came face-to-face with the supernatural entities and events they revere. They would have representations of the supernatural that they learned from surrounding culture or developed themselves, but, absent extraordinary perceptual events, they would have no corresponding direct perceptual experiences of the represented entities.[33] Yet many people want to see, hear, or touch in ways that make the otherworldly objects of reverence *feel* personal—like they've made contact with them (outside whatever mental contact they take themselves to have had).[34] So there is another category of religious credence:

> **Claim 2**: many religious credences *link* particular, concrete entities or spaces from the person's sensible world *to* the supernatural entities described by detached religious credences.

Let's call these *linking* religious credences. Linking takes various forms. This or that particular sensible object might (according to the linking credences) have been <u>touched by</u> a supernatural or otherwise holy being (e.g., the Shroud of Turin, Moses's Staff), something might <u>represent</u> such a being (e.g., Egyptian tomb paintings represent beings that will serve the pharaoh in the afterlife), a certain object might *just be* the supernatural entity (e.g., the host in Catholic mass), and often linking involves a curious oscillation between representation and identification (e.g., the metallic plates under the cloth: *Were* they the golden plates given by Moroni, or did they merely *represent* them?). The exact nature of the linking is often mysterious[35], but in any case, the devoted have the religious credence that the linked sensible entity somehow partakes of the essence or power of the superordinary entity to which it is linked.[36]

Now we can say where sacred values come in:

> **Claim 3**: if a person's linking religious credences designate some specific, concrete entity in the tangible world as linked to one of the supernatural entities described by detached religious credences, then that person's sacred values system attaches to that specific concrete entity.

In other words, linking religious credences—among other functions—deliver entities under their sacralized description *to* the sacred values system, such that *it* can motivate behavior in the physical world in relation to *those* concrete entities. This, then, is what it is for the sacred values system to be *attached*: the person with the relevant sacred values can now find (or produce) entities, actions, events, or places that they will now treat as inviolable in the specified senses; they will also sacralize contact with those entities and regard their value as part of their identity.

Claims 1 through 3 encapsulate the relation between religious credences and sacred values. But we are not quite ready to solve our two puzzles. That's where the following point comes in: entities that are sacralized (a piece of wood, a patch of sand) *gain* a sacred description through religious credences, but they do not lose whatever mundane (factually believed) description they have in people's minds. This gives us:

> **Claim 4**: people with linking religious credences continue to have mostly accurate factual beliefs about the linked entities from the sensible world; such factual beliefs largely correspond to those they would have had about those sensible entities had they not had the linking religious credences.

People do not forget that tomb paintings are motionless paint on a wall; people do not forget that the host is made of grain, which is why Catholic people with celiac disease have difficulty with the sacrament and *know* that they do[37]; people do not forget that the motionless corpse, whose spirit is supposed to be floating nearby, cannot actually hear[38]; and so on. Again, the factual beliefs people have about linked sensible entities persist even as those entities are, *via* linking religious credences, linked to supernatural ones.

Furthermore, for linking and hence sacralization to work at all, such factual beliefs *have to* persist; otherwise, people would be unable to identify *which* entities in the sensible world were the sacred ones. It would do a person no good to have sacralized a certain sensible object through linking religious credences if those linking religious credences—which describe the object as otherworldly—undermined that person's ability to find and identify the sensible object in question; and finding and identification rely on one's having accurate factual beliefs. When one asks oneself or another, "Which entity (of all the entities nearby) is the sacred one?" the answer, in order to be useful, will have to be one that makes the entity in question perceptually identifiable, and such answers typically involve information encoded in routine factual beliefs.

Think about this point in relation to the practices of The Playground described in the Prologue. There, the sacralized items were dolls, sand-castles, and the special space below the wooden playground structure. To be able to locate those sacred entities, the devoted kids had to retain their knowledge (in the form of factual beliefs) that those dolls were made of immobile plastic, that the sandcastles were piles of sand, and that the sacred place was in a schoolyard, for if they had expected, say, the doll/superagents to move of their own accord (as their religious credences described them), they would have been confused about the very dolls they were sacralizing. And so on, mutatis mutandis, for the other sacralized items.

Here, then, is the tension that runs through so much sacred practice and explains why it is characteristically a form of representational pretend play: to treat a sacralized entity *as* what it is described as being by religious credences, one must treat it as having properties that are *other* than what is factually believed about the entity—while *still* (continual reality tracking) having enough of a grip on those factual beliefs to identify and manipulate the object effectively in the physical world. One must then—for one has no other option—*represent* the religiously creeded sacred properties of the entities in question through *symbolic* actions. That is, one must engage in make-believe play. And just as in make-believe play one must retain knowledge of the physical nature of the props in order to move them effectively during the pretense, one must also retain one's factual beliefs about the mundane properties of the props used in symbolic action. The two-map cognitive structure never goes away, because it can't.

We can now solve our two puzzles, starting with the second: *If sacred values are so important, and if sacred entities are inviolable, why do many of the entities to which the sacred values system attaches so easily get discarded or replaced?*

The solution here is straightforward. Since the devoted actor never loses awareness (in the form of factual beliefs) of the mundane properties of the sacralized object (Claim 4), even while having linking religious credences concerning its supernatural properties (Claim 2), she is always aware that the sacred entity—about which she has detached religious credences—is not actually the thing before her: it is in another realm. Thus, especially given the compartmentalization of religious credences, it is always possible for the formerly sacralized entities to *revert* to their mundane, factually believed description, even while the devoted actor maintains adherence to the religion—that is, even while she maintains all the same *detached* religious credences (Claim 1). Linking religious credences are fragile because they are arbitrary and optional, as we saw with the metal plates. True, *while* a linking religious credence is in place concerning a physical object, that object will be treated as inviolable, but that doesn't mean that the linking can't be dropped. This explains the seeming paradoxicality of sacralized entities: inviolable yet discardable. They are inviolable under the superordinary description given to them by linking religious credences, and yet when the links weaken (in Boyer's terms, the sacred object no longer "works"), that description ceases to apply, at which point the formally sacralized object can be discarded. Where, we might ask, are all the slivers of the cross that were so adored in centuries past? Where are the "golden plates" that stayed hidden under the cloth? What happened to the rock in the fork of the tree that had the power to delay the setting of the sun? No one knows, and very few people seem to care.

Now let's turn to the second puzzle: *If sacred values are so important, why do devoted actors often shun or ignore clear instrumental means toward their stated sacred aims?* The answer is that often such instrumental behavior only impacts the prop item under its mundane (factually believed) description; it does not constitute the sort of symbolic action that represents the sacred entities that are the genuine targets of devotion. Sacred action is symbolic, so it is crucial for the superordinary entities *to be represented* through it. And certain action types conventionally constitute representations of such sacred entities, while other action types do not.[39] In cases where the actions do not, it is as if the otherwise sacralized concrete entity were *merely* a prop—and nothing more. Just as a doll on the stage is treated as a precious baby when symbolic action is going on—but is merely an object to be ignored after the play is over—so, too, are embryos in the eyes of the American Evangelical pro-lifers: utterly sacred and inviolable when representational sacred make-believe is occurring, but mere props otherwise. That is why the devoted actors don't support many of the actually effective policies that would reduce abortions: such policies, like affordable childcare, lack the relevant imagined significance and hence aren't sacred.

6. SACRED VALUES: WHAT WE HAVE
MANAGED TO EXPLAIN

These connections to sacred values illuminate what is distinctive about religious credence as a secondary cognitive attitude. They also effectively complete my theory *of* religious credence, which, if we assume an understanding of the technical notions developed thus far, we can now state in a condensed way. Consider:

1) Sam <u>factually believes</u> *that p.*
2) Sam <u>religiously creeds</u> *that p.*

Echoing Hume, we again ask: *Wherein consists the difference?* The first was spelled out in Chapter 2, but we can now add to it. The first means that Sam has a cognitive attitude toward *p* that (in the defined senses) is involuntary, is practical-setting-independent, has widespread cognitive governance, is evidentially vulnerable, *also* plays the Mundane Explanatory Role in action guidance, and characteristically guides actions in concert with the motivation of the utilitarian values system. The second means that Sam has a cognitive attitude toward *p* that (in the defined senses) is voluntary, guides action specifically in sacralized settings, has only limited cognitive governance (is inferentially curtailed), is not evidentially vulnerable, *also* plays the roles of Groupish belief in constituting group identity, and attaches a person's sacred values system *to* real or imagined entities that *p* designates. In sum: both cognitive attitudes exist and are widespread among humans; they differ from each other cognitively in the same ways that factual beliefs differ from other secondary cognitive attitudes, like fictional imagining; and they function in concert with distinct motivational systems. The contrast is clear, and the theory that fleshes out the Distinct Attitudes Thesis and the Imagination Thesis, which have been with us since the Prologue, is now complete.

*   *   *

There is something else to take away from this chapter and the last, which is more external. What has emerged is a preliminary action theory of symbolic sacred action: symbolic sacred action is representational make-believe play, the execution of which is backed by the imperative force of sacred values. That which is sacralized is that which is a prop in such symbolic action, when that prop, according to the conventions of the play, is somehow linked to a revered supernatural entity. Euthyphro should have just told Socrates that that which is holy is merely that which falls into a special subset of such props. It would take some further work to define the general characteristics of that special subset that makes its items holy, but we can easily give examples: copies of the KJV, the plates under the cloth, unborn embryos, and certain patches of land in Palestine. It would be nice if we could ask Socrates what he would say to that, but, unfortunately, there is not going to be a reply.[40]

CHAPTER EIGHT

The Puzzle of Religious Rationality

There is a different way of looking at what I have done so far in this book. I have presented a different *kind* of solution to a problem that is endemic to thought about religious psychology, a problem that lurks throughout the research, even if it is not explicitly acknowledged—from James and Durkheim to Dennett and Barrett.

This is The Puzzle of Religious Rationality. Unfortunately, there is no way of stating it without seeming irreverent, without "breaking the spell," as Dennett puts it.[1] Yet having come this far, we would be intellectually derelict not to consider this issue: it will shed important light on what may be accomplished by a theory of religious "belief."

The Puzzle is as follows. Humans are impressively rational creatures: almost any neurotypical human has the *capacity* to learn to use a computer, learn to drive a car, learn basic facts about how music works, learn to use money, learn arithmetic, learn to use calendars and clocks, learn about a large range of plants and animals, and so on. None of that would be possible without impressive levels of rationality. And here, I mean *epistemic* rationality, which (at a minimum) is the ability to learn and form accurate cognitions in ways that respect evidence and logical coherence.[2] But most humans also have religious beliefs, and many of the contents of religious "beliefs" seem to be anything *but* rational.

Let's review some examples. Dan Sperber tells us that Dorze Christians in Ethiopia "believe" that leopards, which according to them are Christian animals, fast on Christian fast days.[3] So the Dorze maintain that large predatory cats deliberately avoid eating on certain days, because those cats are Christian and that is what Christianity requires. This is, to put it bluntly, contrary to everything people (including the Dorze) know about leopards. E. E. Evans-Pritchard tells us that the Azande "believe" that they can find out about future tragedies or whether a witch was harming someone by poisoning a chicken and seeing if it dies.[4] Such a belief is only supported if you count the Oracle's successes and ignore its failures, and it is utterly mystifying how facts about the future or witches are supposed to causally influence whether the chicken lives or dies. Many mythological religions throughout history have delivered "beliefs" in florid pantheons and miraculous amalgam subdeities—from the elephant-headed Ganesh to the wing-footed Hermes to the pregnant Virgin. All that is the tip of the iceberg of religious belief contents that appear to flout reason.

Imagine that your next-door neighbor started professing "beliefs" with contents like these: *her cat has taken to fasting on certain sacred days, she tells the future by poisoning a parakeet, she knows immortal beings who have human-mixed-with-animal bodies,* and *her granddaughter is pregnant without ever having had sex.* You would be concerned that she had lost her mind. And yet, despite having "beliefs" with analogous contents, the vast majority of religious people appear capable of going about life in a rational way—learning to navigate the world, use tools, exchange goods for money, and so on.

So there is a tension between the apparent irrationality of religious "beliefs" and the rationality of the humans who hold them. The Puzzle of Religious Rationality is to figure out how to resolve the tension in a way that is consistent with the available evidence.

This tension, even when it is not acknowledged in these terms, is intellectually vexing. Kierkegaard senses it, which is partly why he says that the leap of faith one must make to enter into an "absolute relation to the absolute" requires suspending rational ethical thought and universal reasoning.[5] That is a circuitous way of saying that, whatever one gets out of religious "belief," it cannot be achieved through evidence-based rational thought. Yet Kierkegaard was both highly rational and a religious believer.[6] Jason Stigall, a former graduate student of mine who came to philosophy after being a seminarian, formulates a related problem in this incoherent triad in his MA thesis:

(i) Faith is held regardless of evidence.
(ii) Beliefs are based on evidence.
(iii) Faith entails belief.[7]

On reading that triad, many will be tempted simply to dismiss (ii) and pretend to have solved the problem. But that maneuver is sophomoric: it fails to take seriously the impressive rational capacities that even the most ordinary humans have.

Much theorizing in philosophy and psychology of religion can be read as attempting to solve The Puzzle, even when that is not the stated aim. What follows, then, is a taxonomy of nine would-be solution types, which fall under three broader approaches. My aim in this chapter is to argue that eight of these nine solutions do not succeed in solving The Puzzle in an entirely general way. That leaves us in need of another solution: number nine. There will still be moves and countermoves to be made once the chapter is done; proponents of other solution types will wish to plead that the case is not closed. But the evidence by then will not tilt in their favor. Furthermore, my more general aim is to provide a new way of charting the intellectual space in which theories of religious belief occur, in addition to showing that a certain portion of that space deserves exploration of the sort that occurs in this book as a whole.

Here's an overview of the three broad approaches to solving The Puzzle of Religious Rationality, each of which encompasses three solution types.

The first approach takes the contents of religious "beliefs" at face value and attempts to adjust the *apparent* rationality or irrationality of religious beliefs or believers, arguing that these are more or less rational than they seem (and thereby eliminating the tension). Call this the Adjust the Rationality Approach. A second approach involves regarding the contents of religious "beliefs" as *different* from what they appear to be (three solution types fall into this category as well), which allows us to eliminate the tension in a different way. Call this the Adjust the Content Approach. The third is the underexplored Attitude Approach. It focuses on the dimension of mental states that other solutions overlook: attitude type. Importantly, the theoretical virtues of this general approach can be illuminated independently from the particular details of my theory. In any case, three distinct solutions that can be weighed against each other also fall under this approach, with mine being just one—the other two being the "weak belief" solution and Dennett's "belief in belief" solution, both of which I discuss.

To preview the point we'll come to, my solution within the Attitude Approach solves The Puzzle as follows:

> **The Distinct Attitude Solution**: the cognitive attitude of ordinary, every-day (factual) belief is—in point of fact and characteristically in humans in general—constrained by mostly rational processing. But since there is an attitude of religious "belief"—a way of relating to ideas—that is *different in type* from factual belief, the rationality of an agent in relation to her everyday factual beliefs does *not* imply that the contents of her religious "beliefs" can be expected to rationally cohere with her evidence or be internally coherent at all.[8]

Analogously, the rationality of factual belief does not predict that the contents of one's <u>imaginings</u> will be arrived at rationally, since imagining is a different cognitive attitude with different constraints.[9] If that analogy is right, the tension disappears.

One important point before proceeding. In arguing that the other approaches do not solve The Puzzle, I am not saying that *all* their theoretical claims should be rejected (though some of them should be). As you'll see, I think a number of the states and processes posited by the various theorists have a lot going for them. So in arguing that the other solution types don't solve The Puzzle, my main aim is to show that something crucial is still missing from them, which is why the Attitude Approach is needed.[10]

The Adjust the Rationality Approach

Solutions under this approach start by taking the apparent contents of religious beliefs at face value: if people of a certain religion utter sentence S that has literal content *p*, advocates in this family of views accept that those people straightforwardly think that *p*. If someone of Nahuatl provenance says, for example, "There is a feathered snake god" (mutatis mutandis for translation), then that person just takes that literally to be the case. Note that this family of views tacitly consists of One-Map Theories since they don't acknowledge the possibility of a two-map structure, such as I advocate in the previous chapters. That leaves the proponent of any solution under this approach with an especially tough version of The Puzzle since a literalistic attribution of religious "belief" contents leaves one positing beliefs that, on their face, would be extremely hard to arrive at by rational means (like *that a feathered snake god exists*). These solution types are especially pressed to say things that would make adjustments to the apparent rationality of religious believers, to the apparent irrationality of religious beliefs, or to both. Otherwise, the tension will persist.

Solution 1: Religious belief as delusion. This solution says that religious beliefs just are delusions. Strident atheists, like Richard Dawkins and Sam Harris, often suggest such a view, and Dawkins's book is even titled *The God Delusion.*[11] But the view has a longer history than that. Sigmund Freud writes, "Religion would thus be the universal obsessional neurosis of humanity."[12] And in the more recent cognitive neuropsychiatry of religion, Ryan McKay has used the prominent "two-factor model" of delusions to argue for this conclusion: "In the absence of compelling objective evidence for God's existence . . . religious belief is, alethically speaking, pathological and, by two-factor standards, delusional."[13] (The idea behind the two-factor model of delusion is that having delusions requires *both* abnormal experience *and* shortcomings in reasoning capacities; McKay's argument posits both sorts of factors in the origins of religious belief.[14])

All proponents of this solution maintain that religious believers are having delusions. Of course, there's no way to have delusions without being delusional, so the apparent irrationality of religious beliefs is due to deep irrationality on the part of the person who has them. Effectively, then, proponents of this view bite the bullet when it comes to attributing irrationality to billions of people. The Dorze, on this view, would literally think that Christian leopards fast, because they are irrationally deluded to that effect. People who practice ancestor worship deludedly think that invisible deceased ancestors are roaming the forest, and so on.

This solution, more abstractly, alleviates the tension between (i) the apparent rationality of religious people and (ii) the apparent irrationality of religious belief by denying or at least seriously diminishing (i). That's a beautifully simple solution. But it's wildly false: the vast majority of religious people are sane, and most of them lack the afflictions that typically plague the personal lives of people who have clinical delusions.

Furthermore, much research indicates a positive correlation between being religious and having better mental and physical health.[15] Such research does not, in my view, establish that religious "belief" itself is a *cause* of mental health. Perhaps the measured quality of religious people's mental health is a product of having a community of any sort, or perhaps the mental health is itself a cause of seeking community in general, which would include religious community. But such research does indeed refute the idea that religious people are typically delusional.[16]

Solution 2: People are gullible. When I was young, I "learned" a lot about nutrition from a relative. I've since discovered that much of what I learned was false. The gullibility solution, then, says that religious people acquire beliefs about outlandish deities and miracles in much the same way I acquired my outlandish beliefs about nutrition: the pipeline to belief is trust in testimony from a perceived authority, and the apparent irrationality of the beliefs gets explained away by the fact that learning from authorities is generally a good strategy, even if it makes one gullible on certain topics.

This approach appears in Neil Levy's work as a general explanation for apparently irrational religious and political beliefs.[17] Levy thinks that humans do well to trust in authorities who appear *competent* and *benevolent* to them, that we have psychological mechanisms that incline us to trust such people, and that people with outlandish-seeming beliefs are just "epistemically unlucky" in terms of who their perceived competent and benevolent authorities are. This solution preserves the rationality of religious people by saying that it is in general rational to trust competent- and benevolent-seeming authorities, and then it excuses the apparent irrationality of religious beliefs by having them be transmitted by said authorities. Otherwise put, the solution explains the acquisition of apparently irrational beliefs as a by-product of the mostly epistemically useful process of acquiring beliefs directly from reports of authorities.

This is a viable approach at first glance, and it may well apply to many (but not all) people's false beliefs about, for example, climate change: many consumers of climate change–denying right-wing media may *simply be deceived* by the information sources they regard as competent and benevolent (though, of course, others may be in states of ideological credence, as pointed out in Chapter 1).

There are, however, three major problems with this solution as applied to religious "beliefs."

First, this would-be solution diminishes the rationality of people's information consumption too much. More particularly, it sits ill with a large body of research that implies that people, already from childhood, learn *critically* from testimony. As Paul Harris shows through discussion of carefully sequenced experiments, children take note of who has been accurate in the past to determine whom to believe about *new* topics. Children also compare what they hear through testimony with their own experiences to form a measured response to new sources of information.[18] So we cannot posit *uncritical* credulity to explain outlandish religious beliefs. While many people might not know enough climate science to critically weigh their climate change–denying information sources—though that, too, is in dispute[19]—almost everyone from a young age knows that deceased people can't see, that snakes don't grow feathers, that dead people don't come back to life, that water doesn't turn to wine, and so on. So although Levy's testimony-centered explanation of apparently irrational belief might work for a large class of climate change deniers, it fares poorly for religious beliefs, given that religious believers already have background knowledge that should make doubt them the fantastical things that are told of in religious myth and doctrine. The level of credulity this approach posits diminishes the rationality and critical thinking skills of religious people too much—well below the level of critical reflection on testimony that people demonstrate even as children.[20]

Consider this: it is easy for most people to see that *other people*'s religious beliefs sit ill with basic facts about the world, which is why people generally don't just adopt all religions they encounter, no matter how competent and benevolent the proselytizers are. So people certainly have the *capacity* to detect that their own community's religious beliefs *also* sit ill with their background knowledge of basic facts about the world—as the developmental psychology just canvassed would imply. So why wouldn't they exhibit the sort of skepticism that is characteristically linked to that capacity when it comes to their *own* religion? Levy's gullibility approach fails to answer that question, which is a stroke against it.

The second problem with this solution is that it generates a further puzzle as to why people are so doggedly resistant to *changing* religious beliefs. When it came to my own false beliefs about nutrition, I updated readily and easily when I got better information (and I discarded the idea that my relative was competent on this issue). So if religious belief acquisition was just like *that,* religious beliefs should update just as easily. After all, people don't like to be fooled or stay fooled. But with religious beliefs, evidence-based updating is more the exception than the rule, as we've seen. Recall that cult members often become *more* entrenched in their beliefs when they receive contrary information, which cannot be explained by the idea that they simply had a bad original source[21]—or were just "epistemically unlucky," as Levy would put it.

The third problem is that the gullibility solution makes religious belief acquisition all too passive, which doesn't cohere with the fact that people have to *work hard* to form and maintain their religious beliefs, as we've seen.[22] On Levy's picture, people catch their beliefs as easily as I apparently caught false beliefs about nutrition; I just absorbed the information, and it stuck (at least until I learned more). But if that's how it worked, why would religious people in so many traditions "struggle to believe" and "wrestle with faith"—not to mention performing lengthy rituals designed to maintain their beliefs?

Thus, the idea of naive acceptance of testimony—gullibility—falters at solving The Puzzle for three related reasons: (i) it diminishes the rationality of how humans process testimony too much—it thus sits ill with what we know of *critical* trust in testimony from developmental psychology, (ii) it leaves us with a further puzzle of why people would be so resistant to changing religious beliefs when contrary evidence emerges, and (iii) it fails to capture the active hard work that people put into forming and maintaining religious beliefs.

Solution 3: Religious belief as rational. This is the route of religious apologists. The view here is that, despite being surprising, evidence does favor the contents of religious beliefs. In other words, this approach eliminates the tension that The Puzzle highlights by saying the apparent irrationality of religious beliefs is *mere* appearance. This is the route from Anselm to Plantinga, with many other ardent efforts in between. I have very little to say about this type of solution for the following reason: if it were successful with respect to any one cluster of religious beliefs (which I find unlikely), it would still leave The Puzzle of Religious Rationality in place for all or almost all the others around the world. Suppose, for example, that St. Anselm's apologetics in his later work *Cur Deus Homo* (*Why God*

Became Man) were successful at demonstrating the rationality of Christian belief in the incarnation of God.[23] That would still do nothing to alleviate the apparent irrationality of belief in Hermes, Quetzalcoatl, or Ganesh—or in invisible deceased ancestors or the predictive powers of chicken poisoning. We'd still have The Puzzle of Religious Rationality on our hands in spades, only having alleviated it for people in one religious tradition (if that).

* * *

All three of these solutions tacitly assume a general One-Map Theory: they posit one kind of belief state that is common to everyday cognition and religious cognition, and then they attempt in various ways to reconcile the apparent rationality of the former with the apparent irrationality of the latter by diminishing those appearances of rationality or irrationality one way or another. But for each solution type, the attempt is unlikely to succeed.

I have already detailed reasons for rejecting One-Map Theories in relation to specific communities like the Vineyard (Chapter 3), and I've laid out crosscultural evidence that suggests that One-Map Theories are unlikely to be true in a wide range of cultures around the world (Chapter 4). The discussion here, however, provides a more general reason for being skeptical of One-Map Theories: the attempts at solving The Puzzle of Religious Rationality that tacitly presuppose a One-Map Theory fall short.[24]

The Adjust the Content Approach

The next set of solutions alleviates the tension between the general rationality of humans, including religious ones, and the apparent irrationality of religious beliefs by saying that the actual contents of religious beliefs aren't what they appear to be.

To judge from their expressions in language, religious beliefs are about supernatural deities, ancestors, spirits, magic, and so on. But, say solutions under this approach, such linguistic expressions should *not* be taken at face value, and what religious beliefs are actually about is something else—or nothing at all. On the previous set of solutions, which held contents fixed at face value, the tension inherent in The Puzzle was alleviated by adjusting posited levels of *rationality* (real or apparent) on the part of either the religious beliefs or religious believers. The present set of solutions leaves in place the rationality of religious believers and the irrationality of apparent religious belief contents, and it attempts to resolve the tension by saying that the *actual contents* of religious beliefs are not the irrational ones that would be ascribed on a superficial examination of their linguistic expression. Religious people, on these approaches, believe not as they speak.

Solution 4: Displaced content. This solution eliminates the apparent irrationality of religious beliefs by saying that, despite appearances, they have contents that concern something altogether different from the supernatural entities that their linguistic expressions (in myths and doctrines) seem to describe. There are two major versions of this solution: one version says that religious beliefs have moral contents; the other says that religious beliefs are about the believer's society or group.

Stephen Jay Gould's idea of *non-overlapping magisteria* (NOMA) is an example of the moral content solution.[25] Gould claims that science and religion don't conflict, because they deal with different "domains." Religions are about the moral domain, and since moral principles are not what science is about, there isn't a conflict between science and religion. Such a view feels attractive since it seems to promise to circumvent ongoing culture wars. Unfortunately, Gould's view paints a distorted picture of what many religions are actually like.

The main problem with NOMA is that most religions contain *many* descriptive claims about what the world is like—in addition to moral claims. Such descriptive claims are believed in some sense, and it is implausible that *all* of the apparently descriptive beliefs can be reinterpreted in a moral way. What, for example, are the moral contents of the beliefs *that Jesus turned water into wine* or *that Vishnu became incarnate as Rama*? Perhaps there are moral *implications* of those beliefs, given background ideas one might have, but to work out those implications, one first has to take the apparent contents seriously, including those that accord Jesus and Rama existence and supernatural powers. To make matters worse for Gould, many of the descriptive contents of religious beliefs are ones on which scientific reasoning and evidence can bear. The closure of the physical, for example, rules out the causal efficacy of incorporeal spirit agents, which is a central posit of religious beliefs in many traditions, such as animism, many forms of Buddhism, ancestor worship (which is common globally), and most varieties of the Abrahamic faiths.

Gould's approach, interpreted as charitably as possible, requires us to look at any cluster of religious descriptive claims as being one long parable that somehow expresses only underlying moral beliefs. This may be an attractive way of thinking about religious myth and doctrine for someone who, like Gould, is nonreligious. But it does not capture the mindset of those religious people who take the descriptive contents of their religious beliefs seriously and even say as much. To vary the theme from the end of Chapter 6, Gould's approach lacks the expressive power to distinguish the actual religious believer from the appreciative atheist who is of like moral mind and enjoys stories. In addition, religious schisms and even accusations of heresy often occur over differences in descriptive claims that one sect or another makes, without there being obvious independent differences in the basic moral views of the respective groups. Such schisms would make little sense if the respective groups were *merely* choosing different parables for the same underlying morality. So the descriptive contents of religious myths and doctrines cannot be neatly set aside as mere parable, and many people—*pace* Gould—do in some relevant sense "believe" those descriptive contents.

This brings us to the other major version of the displaced content solution. Call this the "clannish content" version.

Émile Durkheim, the chief proponent of clannish content, actually comes close to stating The Puzzle of Religious Rationality—and perhaps even does so, if you interpret the quotations below in a certain way. In a fairly early passage of his seminal work *The Elementary Forms of Religious Life*, he writes:

> Now, the religious representation of the universe was too crudely truncated, especially in the beginning, to have fostered practices useful in daily life. It views things as nothing less than living and thinking beings—consciousness and personalities like those the religious imagination has made the agents of cosmic phenomena. It was not by conceiving of them in this form and treating them accordingly that man made them serve his ends. It was not by praying to them, celebrating them with feast days and sacrifices, with self-imposed fasts and privations, that he could stop them from harming him or forced them to further his plans. Such procedures could have succeeded only very rarely and, to say the least, miraculously. If

religion had to be justified by giving us a representation of the world that would guide us in our dealings with it, religion could not have performed this function, and people would have been quick to notice. *The failures, infinitely more frequent than the successes, would soon have warned them they were on the wrong track, and religion, given the lie at every turn, could not have endured.*[26]

So far, Durkheim and I are on the same page. Much of what he says in this passage can be read as a notational variant of ideas I advocated in Chapter 6: religious beliefs are ill-suited to play the Mundane Explanatory Role in relation to behavior since, in playing that role, they would have led to continued action failure. But most humans, he implies, would not persist in relying on beliefs of any sort in a way that led to continued action failure, and by implying that, he essentially decides on an approach to The Puzzle that affirms the general rationality of religious *people*. As a result, Durkheim decides that the apparent irrationality of religious belief must be dealt with somehow. So how does he do it?

I quote again at length, this time from the middle of the book, where he is criticizing previous theories as well as advancing his own solution:

> And in order to explain how the notion of the sacred could emerge . . . most theorists were forced to assume that man superimposed an unreal world on the reality he observed. This world was said to be constructed entirely of fantastic dream images or monstrous aberrations which the mythological imagination invented under the marvellous but deceptive influence of language. But if so, it is impossible to understand why humanity should persist for centuries in the errors that experience must have quickly exposed.
>
> Adopting our point of view, these difficulties disappear. Religion is no longer some inexplicable hallucination and becomes rooted in reality. We can say that the worshipper is not deluding himself when he believes in the existence of a higher moral power from which he derives his best self: that power exists, and it is society. When the Australian is transported beyond himself and feels life flowing in him with an intensity that surprises him, he is not prey to illusion. This exaltation is real, and it is really the product of forces external and superior to the individual. Of course he is mistaken when he believes that this heightened vitality is the work of a power that takes plant or animal form. But his error lies only in taking literally the

symbol that represents this being to men's minds, or the form of its existence. Behind these figures and metaphors, crude or refined, there is a
concrete and living reality.

   Religion takes on a meaning and a logic that the most intransigent ra-
tionalist cannot fail to recognize. The main purpose of religion is not to
provide a representation of the natural world, for if that were its basic task
its persistence would be incomprehensible. In this respect it is scarcely
more than a tissue of lies. But religion is above all a system of notions by
which individuals imagine the society to which they belong and their obscure yet intimate relations with that society. This is its primordial role;
and although this representation is metaphorical and symbolic, it is not
inaccurate. Quite the contrary, it fully expresses the most essential aspect
of the relations between the individual and society.[27]

This passage admits of multiple interpretations, though one thing is clear
no matter what the interpretation: though he formulates it differently,
Durkheim has set forth a solution to what I call The Puzzle of Religious
Rationality.

   The reason the passage admits of multiple interpretations is that Durkheim seems to both deny and affirm that there is a layer of representations of the supernatural in the mind of the religious adherent. He seems
to deny it when he writes disparagingly about theorists who were forced
to assume "that man superimposed an unreal world on the reality he observed." If such a posit is as absurd as Durkheim implies, then there must
be no map layer representing the supernatural. Yet he seems to affirm that
very thing when he writes about how the religious people he discusses
commit the "error" of "taking literally the symbol" that stands for society
and social relations: if those people take the symbol literally, then there
must be a representation of the supernatural after all. So it is unclear
whether Durkheim posits a supernatural layer of belief or not. But his
overall message *is* clear either way: religious beliefs are *in fact* ultimately
about *society,* despite surface appearances, which makes them at the end
of the day mostly true and accurate since society is something that exists.
The apparent supernatural content of religious beliefs is displaced by clannish content.

   Given what I said in the last two chapters, it's clear that I think Durkheim is on the right track with the idea that religious beliefs are typically

clannish in some way. Nevertheless, Durkheim seems to be guilty of a mistake that must be corrected. We must differentiate two things:

(1) The content of religious beliefs (that is, what they describe as being the case).
(2) The social function that religious beliefs play.

Durkheim is entirely right about (2) since religious beliefs do play pervasive roles in constituting group identities. But that does not imply that their *contents* (1) are about the social group to which one belongs. Since he infers social content (accuracy conditions) from social function, Durkheim, I think, is guilty of a gross non sequitur.

His mistake becomes obvious when we consider religious narratives that are more elaborate than the representations of simple totem animals. Consider the Jonah story. That story has detailed contents, which we would be hard-pressed to interpret as being about a given society and the relations between people that exist in it. What social relations are represented by sentences that appear to claim that a man lived in the belly of a whale for three days, which then belched him up? Perhaps there is some moral lesson of social import, but this social import is murky at best, while the actual story contents are clear as day.

The prospects for displacing *all* the apparent contents of religious beliefs with clannish contents are thus dim. So Durkheim, for all the merits of his theory, does not provide a general solution to The Puzzle of Religious Rationality.

Importantly, Gould and Durkheim end up having the same thorn in the side of their proposed solutions: religious beliefs with *descriptive* supernatural contents that are both clear in themselves and hard to interpret as metaphors for something else. Though *some* classes of religious beliefs might be amenable to reinterpretation with displaced contents, that large class is not. And, importantly, it is also the very class that drives The Puzzle in the first place.

Solution 5: Murky contents. Another solution under the Adjust the Content Approach is the idea that the contents of religious beliefs are too *murky,* or unclear, for them to be subject to decisive rational scrutiny. As it happens, I think a view like this applies to *many* religious beliefs. But the question is whether it applies to enough of them to make The Puzzle go away.

TABLE 8.1

	Sperber (1975)	Sperber (1982)	Sperber (1997)
Row 1	Symbolic representations	Representational beliefs	Reflective beliefs
Row 2	Conceptual representations	Factual beliefs	Intuitive beliefs

Dan Sperber's famous theory of representational beliefs arguably falls under this solution type. Sperber introduces this theory under different terminologies over time, but the basic contrast is more or less the same. Table 8.1 shows the different terminologies that Sperber uses for his central distinction. Terms in Row 1 are (more or less) notational variants, as are terms in Row 2. For clarity, I deploy Sperber's 1982 terminology, though nothing important hangs on that, and I use his other writings for supplemental help with understanding the central distinction.[28]

Sperber's basic idea is that some beliefs have a representational structure that involves a quotation, where what's *inside* the quotation is often not fully understood by the person who holds the belief. An example of this would be most people's belief in the Trinity. Most people who profess to believe that "God is three persons in one" do not (by their own admission) understand what that sentence means. So how can they believe it? Sperber says that the internal representational format of their belief is something like this:

the Church authorities say, "God is three persons in one"

That is, the representational structure of the belief includes an internally quoted sentence (or other quoted representation), where that sentence is not fully (or perhaps not at all) understood. Thus, this is a representational belief because it is a belief in some sense *about* a representation (i.e., about the bit inside the quotation). Nevertheless, on Sperber's view, the quoted sentence is somehow endorsed by the "validating context," which attributes that sentence to some authority figure (or gives some other form of approval).[29] In this case, the validating context is this: *the Church authorities say_____.* Then, by virtue of the believer's allegiance to what the Church authorities say, she may be said to representationally believe that "God is three persons in one," despite not having a clear-cut proposition in mind by which to interpret that sentence. To put it crudely,

representational beliefs are often those in which a person internally ver-
balizes a certain sentence—and perhaps even mouths it to themselves—
and despite not knowing exactly what it means, has an approving attitude toward it by virtue of the connected validating context.

Sperber's contrast of interest is between ordinary factual beliefs (ones
that aren't representational) and representational beliefs (of the sort just
described). Sperber sees ordinary factual beliefs similarly to the way I do,
which is why I quoted him approvingly in earlier chapters. As he puts it,
factual beliefs are stored in "encyclopedic memory" and subject to "strong
criteria of rationality."[30] The key is that *ordinary* factual beliefs, for him,
do not have the quotational structure of representational beliefs. One just
believes the understood contents directly. For example, a factual belief for
Sperber might have contents like these:

> *the church building has a parking lot*

Though the person with the relevant factual belief represents this content,
that content is not *about* a representation; it is about a church building
with a parking lot. One can expect factual beliefs like this to behave in a
more or less rational fashion. Since the contents are understood and not
quotational, the beliefs themselves will be easily updated.

An analogy is useful for understanding Sperber's central idea. Suppose
you don't understand the equation $e=mc^2$. Still, you know how to pro-
nounce it; you can repeat it at will by saying it out loud or just in your
head; you know that Einstein, who was a genius, asserted it; and so on.
In this scenario, your lack of understanding of the equation prevents you
from having a direct factual belief with these relevant contents:

> *matter converts to energy at the ratio of the square of the speed of light*

Nevertheless, you are perfectly capable of having a representational belief
with these contents:

> *physicists maintain* "$e=mc^2$"

And even these:

> *it is true that* "$e=mc^2$"

What makes the beliefs here "representational" is that the believing agent has a linguistic representation encoded in the internal belief structure, so the agent thereby representationally believes that representation. For Sperber, then, the Christian's belief that "God is three persons in one" is like one's belief (in this example) that "e=mc²": in both cases, the contents of the quoted portion are not understood by the believer. That lack of understanding makes the representational beliefs less susceptible to rational updating. Of course, for some representational beliefs, like the one about the Trinity and unlike the one about Einstein's equation, it may be that *no one* has a clear understanding of the quoted sentence; hence, those *symbolic* representational beliefs may evade rational scrutiny indefinitely, no matter what the evidence.

Sperber's theory presents a potential solution to The Puzzle. Religious beliefs, like the one about the Trinity, are representational. And since their quotational portion is poorly or not at all understood, they end up having "murky" contents—Sperber calls them "semi-propositional"[31]— that evade confrontation with whatever evidence one encounters. If the believer confronts evidence that goes against one interpretation of the quote inside the representational belief, she concludes that some *other* interpretation must be correct. Thus, despite the *apparent* outlandishness of many religious beliefs, their representational/quotational structure makes it possible for a rational person to representationally believe them without devolving into irrationality. Is this a good solution?

There is one hitch. Sperber does not usually consider the issue of attitude type, so it is not immediately clear from his writings whether he thinks the cognitive attitude that people take *toward* the *whole* representations involved in their representational beliefs (taking the validating context and the quotation together) is the same as, or different from, the cognitive attitude they take toward the representations involved in their factual beliefs. Yet what little he has said indicates that he does *not* think there is a variation in attitude type.[32] Thus, if we simply label as FACTUAL BELIEF the attitude type involved in the belief about the church parking lot, we would describe that belief as follows:

FACTUAL BELIEF: *the church building has a parking lot*

For Sperber, then, the attitude type involved in representational beliefs would be the same as the attitude type involved in ordinary factual beliefs. That means we could describe the representational belief about the Trinity like this:

FACTUAL BELIEF: *the Church authorities say,* "God is three persons in one"

For Sperber, again, the *attitude type* is not different; there is just the difference in representational structure, which makes a corresponding difference in what you can say about their contents. Otherwise put, Sperber tries to get all the explanatory work out of a distinction in representational format, without allowing for variation in cognitive attitude between factual beliefs and representational beliefs. Another way to put this is to say that Sperber is essentially a One-Map Theorist who tries to accomplish many of the same things I do with my Two-Map Theory, but he does so by positing a different type of representational structure only. This lack of differentiation in attitude is exactly where problems start to arise.

It may at first seem to be an advantage of Sperber's theory (in comparison with mine) that it avoids positing a distinct cognitive attitude since it seems more parsimonious.[33] But two things should be noted before drawing that conclusion. First, the parsimony advantage of Sperber's theory is slight at best: we already know that humans are capable of secondary cognitive attitudes, and my posit of religious credence is just another secondary cognitive attitude, so it is not a major addition to our psychological ontology beyond the sort of thing that is already commonly recognized. Second, Sperber's account tacitly requires additional attitudes anyway. Consider again the example about the Trinity, this time with the attitude type noted:

FACTUAL BELIEF: *the Church authorities say,* "God is three persons in one"

Crucially, many *atheists* also have a factual belief of this form. After all, that *is* what the Church authorities say, and many atheists happen to know and hence factually believe that. As it stands, this belief merely tracks what certain people say. Sperber needs to posit an *additional attitude* to explain the allegiance that people have to whatever it is the Church authorities say (though he does not, as far as I can tell, clearly acknowledge that theoretical need). Now, one might posit something like this:

FACTUAL BELIEF: *whatever the Church authorities say is true* which would easily give us:

FACTUAL BELIEF: *it is true that* "God is three persons in one"

The problem with that, however, is that basically no one (including the Church authorities themselves) *actually thinks* that everything they say is true; rather, even the devout tend to *voluntarily* pick and choose from among things that Church authorities say, as discussed in Chapters 3 and 4. So the intermediate posit is implausible, and hence, so is the derivation of the final factual belief. What Sperber needs in his account is an attitude of *allegiance* to certain core doctrines—which are believed as quotational sentences—and not merely the attitude of mundane factual belief. And in needing an attitude of allegiance for his account, Sperber seems to need the very sort of attitude that my account posits, in which case, the claimed parsimony is illusory. Furthermore, this discussion makes clear that Sperber's view, as it stands, will have a hard time making sense of the *voluntariness* of many religious beliefs since the only attitude he ultimately posits is factual belief and factual beliefs are not under voluntary control.

Yet there is a more damning criticism of the theory of representational beliefs, *insofar as* it is intended as a solution to The Puzzle (an intention that does seem to fall within the purview of Sperber's ambition, since his 1982 paper is titled "Apparently Irrational Beliefs"). It's just that *many* religious beliefs do have descriptive contents that are clear enough for rational appraisal. Recall from Chapter 4 that Mormons believe that American Indians descended directly from the ancient Israelites. This is not a difficult-to-interpret semiproposition; it can be assessed in relation to evidence, and it has been. Yet it persists. Doomsday cults have specific beliefs about what day the world will end. Many creationists believe that the world was created in seven days, where "day" is intended to mean a twenty-four-hour period. Young Earth Creationists also "believe" that the world is less than ten thousand years old, and that is certainly clear enough to be subject to rational scrutiny. And consider this passage from Pascal Boyer, which describes common beliefs among the Fang of Central Africa:

Among the Fang some people are said to possess an internal organ called *evur,* which allows them to display particular talent in various undertakings outside the domain of everyday activities. People with great oratory skills or a particular ability in business, people whose plantations are especially successful, are commonly said to have an *evur*. This is usually described as a small additional organ located in the person's stomach. . . . Some *evur*-bearers are said to launch invisible attacks against other people, drink their blood and bring misfortune, illness or even death to the victim.[34]

This cluster of beliefs may be vague in various places, but it affords enough points of clarity that its contents are not impossible to assess relative to the relevant evidence, which at a minimum would include anatomical study.

Many such examples exist of descriptive beliefs in various cultures and religions that are *not* murky enough to evade the possibility of rational scrutiny—whether their internal coding is representational or not. So for *that* class of beliefs, for all Sperber has said, The Puzzle of Religious Rationality persists since, as noted, the people who hold them generally *are* rational. Sperber's theory implies the following conditional: *if the quotation inside a representational belief is well understood, then that representational belief will be subject to the same "strong criteria of rationality" as regular factual beliefs.* And yet for many mooted representational beliefs, the antecedent is true, while the consequent is false.

I do think there is much to Sperber's idea of representational belief, and I do think it goes some way to alleviating The Puzzle. But it doesn't solve the puzzle entirely, and it appears to need the additional element of a different attitude type anyway.

Solution 6: No content. The last solution under the Adjust the Content Approach is the most radical. Yet it has had curious appeal for many theorists of religious belief and ritual over the last century. This appeal is mainly due to its association with Ludwig Wittgenstein's enigmatic yet strangely compelling writings on religion.[35] The general idea—and it is disputed whether Wittgenstein actually thought this[36]—is that religious language, rituals, and even beliefs (those being in this case whatever it is that underlies the language and rituals) *don't have any content at all.* That is, they aren't representations that can even be true or false, accurate or inaccurate. Rather, they are *expressions* in symbolic form of deep experiences, needs, longings, hopes, and feelings. Positions like this go under the heading *noncognitivism* since a cognitive state is one that *can* be true or false, or accurate or inaccurate.

This is not the place to engage in extended Wittgenstein scholarship, since the aim here is to discuss a certain position's potential for solving The Puzzle rather than to figure out Wittgenstein's elusive views. So for the sake of argument, let's grant that Wittgenstein did hold a noncognitivist view and try to assess that view in relation to the present problem.

In his *Remarks on Frazer's Golden Bough,* Wittgenstein comments: "Frazer's account of the magical and religious views of mankind is unsatisfactory; it makes these views look like *errors.*"[37] This suggests that Wittgenstein saw Frazer (rightly or wrongly) as thinking that people who practice rituals have straightforwardly false beliefs about their causal efficacy: just as someone might try to turn on the oven by turning the knob for the stovetop due to a simple false belief, so, too, did Frazer think that premodern people who, say, try to bring a good harvest by making a sacrifice to a certain deity simply have false beliefs about the ritual's causal powers. In other words, they commit *"errors."* But Wittgenstein thought Frazer was wrong about this and proposed an alternative:

> Burning in effigy. Kissing the picture of one's beloved. That is *obviously not* based on the belief that it will have some specific effect on the object which the picture represents. It aims at satisfaction and achieves it. Or rather: it *aims* at nothing at all; we just behave this way and then we feel satisfied.[38]

Just as kissing a photo does not rest on the false belief that the beloved will feel the kiss, neither do rituals rest on false beliefs or naive errors.[39]

These remarks are insightful, but one way of precisifying them leads to the sort of noncognitivism currently under discussion: religious states of any sort are not errors like Frazer thought, *because they are not the sorts of states that can be true or false at all.* That is, they have no representational contents and hence can't have content that is in error. In spelling out this line of interpretation, Hans-Johann Glock writes in his *A Wittgenstein Dictionary*:

> *The non-descriptive and non-cognitive nature of religion* Religious statements do not describe any kind of reality, empirical or transcendent, and do not make any knowledge claim (LC 59–63). Someone who believes in a Last Judgment does not use expressions like "Such and such will happen" to make a prediction, but rather to express a commitment to a "form of life,"

for example one in which people feel constantly admonished by God's approval or disapproval. Indeed, if he were making a prediction, it would not count as a religious belief (LC 56–58; CV 87). Someone who believes in an afterlife is not committed to the Cartesian notion of a soul-substance, but only to a certain picture, although Wittgenstein sometimes admits that he does not have "any clear idea" of what the picture amounts to (LC 70–71; PI II 178; RPP I §586). The belief in miracles is a propensity to be impressed by certain coincidences. Someone who says "It is God's will" may be uttering something similar to a command like "Don't complain!" (CV 61).

   The existence of God By the same token, religious terms like "God" do not refer to entities, and to state that God exists is not to make a statement about the existence of a certain entity (LC 63; CV 50, 82).[40]

If this sort of noncognitivism were to apply to religious "beliefs" generally, it might seem questionable to call the underlying mental states that drive religious behaviors "beliefs." But the position, if true, would solve The Puzzle. On this *no content solution,* otherwise rational religious people would not be guilty of forming extensive collections of beliefs with outlandish and irrational contents, because the mental states in question *have* no contents. And with that, The Puzzle disappears!

   Before I critique this apparent solution, let me make a comparison that will illuminate the psychology it posits. When he was coming up with the lyrics for "Hey Jude," Paul McCartney was troubled by a line that he had felt compelled to include: "The movement you need is on your shoulder." He worried that it didn't *mean* anything and thought that was a reason to strike the line, even though he liked it. But when he presented the line to John Lennon, Lennon said it was fine because it made him *feel* a certain way.[41] With this response from Lennon, McCartney felt better about the line and decided to keep it, as we all know.

   On the noncognitivist/no content line we are currently exploring, religious beliefs *as quasirepresentational structures* are something like internal versions of that McCartney lyric. They may have structures similar to content-bearing representations, but those internal structures don't have any accuracy or truth conditions (just like "The movement you need is on your shoulder" can't really be true or false). They rather have a certain resonance, such that rehearsing them (internally or externally) brings about a certain "satisfaction" that makes them worth holding dear. This gives a strong response, I think, to the critic who would say the noncognitivist psychology is incoherent for positing beliefs without contents: we may quibble over whether to call the content-free internal states "beliefs," but the posited states are possible and may indeed often occur.

Does this solve The Puzzle?

It *would* solve The Puzzle *if* it applied generally (or even mostly) to the class of mental states known as religious beliefs. The problem is that too many of these states do have identifiable contents. To judge from their linguistic expressions, many religious beliefs describe miraculous events that were supposed to have happened in history. Such events may or may not have happened, such that we can reasonably say the contents of the beliefs are true or false. Many religious beliefs posit causal connections, such as the connection between prayer for healing and healing, or between sacrifice and good outcomes (either generally or in particular instances), and such causal connections may or may not exist. Furthermore, even if one wishes to see rituals themselves as mere expressions that achieve a mysterious satisfaction, there are many descriptive religious beliefs that are not connected to rituals, including many theological doctrines, cosmologies, and mythological stories. And claims of truth and falsity ("it is true that . . . ") seem to apply sensibly to many of these beliefs in a way that they do not apply to "The movement you need is on your shoulder." Furthermore, the fact that claims of truth and falsity do sensibly apply explains why so many people throughout history have bothered to *argue* about the truth or falsity of various religious beliefs, even when there is little difference that such arguments would end up making to the downstream rituals that are performed. If one merely aimed at a feeling of satisfaction, why bother arguing?

There is thus compelling reason to think that a large fragment of those mental states we call religious beliefs do have contents. And there are two further points to make in favor of that position.

First, recall from Chapter 4 that many people have *difficulty* holding religious beliefs. "Yes, that *is* hard to believe," says the Anglican future Archbishop of Canterbury to Anthony Appiah's mother when she was a girl. Luhrmann and Boyer, as anthropologists, point out that many religious rituals appear to be designed to put in place or hold in place religious beliefs that are difficult to maintain.[42] And Pascal's Wager is a case for doing what it takes to overcome the *difficulty* of acquiring religious beliefs concerning events such as the resurrection. The most plausible explanation for these difficulties is that the *contents* of the religious beliefs themselves tend to appear highly implausible, which would not be the case if, as the noncognitivist claims, the religious beliefs in question did not have contents.

Second, several of the roles that religious beliefs play in constituting people's group identities depend on their actually having contents. Recall from Chapter 6 that religious beliefs, like other Groupish beliefs, work well at distinguishing one group from another *because* many of them are false or at least implausible. This is what ensures that people outside the group will not hold them, which helps make the in-group a meaningfully distinct cluster of persons. Yet falsity and implausibility are only features of mental states that have contents. Furthermore, recall the important study by Richard Sosis and Candace Alcorta on religious versus secular kibbutzim in Israel.[43] One of their important findings was that the presence of burdensome rules in the religious kibbutzim led to greater longevity of those communities, but this was not the case for secular kibbutzim that lacked religious beliefs. That difference would be puzzling if religious beliefs were merely commitments to a "form of life," as opposed to being mental states with distinctive contents since joining a secular kibbutz with burdensome rules is also a commitment to a form of life. If we assume, however, that the belief "that God exists" has the content it appears to have, we arrive at a strong differentiator between the religious and secular kibbutzim, one that would help explain why the religious ones have more powerful group identities and hence why they last longer as communities and institutions.

Importantly—to conclude this subsection and transition to the next—there is another way to capture Wittgenstein's intuition that religious rituals are not "errors" in the sense of resting on simply false factual beliefs. The approach I am driving at will also help make sense of Gould's intuition that religion and science are in some way "non-overlapping." It holds that religious people have a different *attitude* type toward the contents of their internal religious representations. On this approach, the religious attitude, even when it has false contents, is not one of passive error—as if one merely read the instruction manual incorrectly—but one of an active embrace of the sort Kierkegaard describes. This approach has the further virtue of allowing us to say everything Durkheim wanted to say about how religious beliefs are connected to one's identity group, *without* forcing on us the fancy maneuvering of trying to change their contents from what they appear to be. So if you had the vague feeling that *something* was right about some of the solutions I've described so far, without being fully convinced due to their shortcomings, you are likely to find this next approach attractive.

The Attitude Approach

Recall our framework from Chapter 1 for thinking about mental states, which cast each mental state as consisting of four components: **a**gent, **R**elation/attitude, **m**ental representation, and **c**ontent (**aRmc**, for short). Having a cognitive attitude of any sort involves these four components.[44] Of course, we should apply this framework to religious beliefs. Let the word "believes" in the following mental state description designate an attitude of religious belief, whatever that amounts to:

Andrew <u>believes</u> *that the ancestors desire that he sacrifice an ox.*

The tension that The Puzzle highlights is between Andrew's general rationality and the outlandishness and apparent irrationality of the idea that dead people would want anything, let alone a meat sacrifice (which Andrew and his family and friends will be the ones to eat anyway). Thus far, we have seen attempts to alleviate this tension in various ways. Claiming Andrew (the agent) is delusional puts *his* level of irrationality into accord with the outlandishness of the contents of the belief; arguing that the contents are somehow evidentially supported puts the contents' level of rationality in accord with Andrew's; saying that the contents are really moral, or about the clan, or nonexistent eliminates the tension by making half of it disappear; saying that the belief is *representational* (Sperber's term) makes it primarily about a vague representation and thus does much the same. So the solutions we have canvassed so far have involved positing adjustments to the agent, the mental representation, or the content components of the religious belief mental state. Yet none of the solutions, all of which skipped over attitude type, was satisfactory. We thus have a compelling reason, from a systematic point of view, to suspect that the <u>attitude</u> component of religious belief is the proper locus of a solution to The Puzzle.

The next three solutions take that approach.

Solution 7: Belief in belief. Dan Dennett puts forth the idea that many religious people do not actually believe (by which he means roughly what

I mean by factually believe[45]) their religion's core myths and doctrines; rather, they believe that good things would flow from believing them. And so they try to get themselves to believe them, try as much as possible to act as if they do, and perhaps even try to convince themselves to have the belief that at some level they are aware they don't have. Hence, "belief in belief." This solution falls under the Attitude Approach because that's the dimension of focus on Dennett's view. More formally, let p be some religious doctrine that a religious person (call him William) asserts with apparent sincerity. Dennett is saying that this may not be the right description of him:

> William believes *that p.*

Rather, this would be:

> William believes that good consequences follow from believing *that p.*

In arguing for the lack of first-order belief, Dennett makes a point I've alluded to in earlier chapters: that much of the time, religious people do not behave as if their religious "beliefs" are true:

> If you really believe that your God is watching you and doesn't want you to masturbate, you don't masturbate. (You wouldn't masturbate with your mother watching you! How on earth could you masturbate with God watching you? Do you *really* believe God is watching you? Perhaps not.)[46]

I think this observation is underappreciated. Even the most devout people *often* behave as if there are no supernatural beings who take an interest in their affairs, despite professing otherwise when other *people* are watching. I recall talking to a woman whose family was staunchly Southern Baptist of the "no drinking" variety. She told me that her family nevertheless drank regularly during the week, and when the preacher on Sunday called people to come forward to confess if they'd been drinking, they wouldn't come forward, due to the social shame they would feel. Even in church, they were only concerned with what other *people* would think—never mind *God's* view of the matter. Did they *really* believe that God was present at the church service? They almost certainly didn't factually believe it. If they did, why were they not bothered about perpetrating a deception in front of *Him*? We could, of course, imagine the people in question having rationales for not coming forward that could be made to be consistent with the idea that God was indeed present, but those rationales would be of a freely chosen, improvised character that would further betray that they are not factually believed. Dennett and I are thus in agreement that factual belief in professed religious ideas is often lacking.

Furthermore, Dennett's view goes some way to alleviating the tension inherent in The Puzzle. There is, after all, plenty of evidence that belonging to a religion has various benefits, including social, emotional, and even physical ones. And if you're taken by the idea that belonging to the religion requires "belief," it may be in some sense perfectly rational to "believe in belief"—even if the first-order belief that you believe you're supposed to have would not be rational by evidential lights.

For all its merits, however, Dennett's "belief in belief" view doesn't solve The Puzzle in a *general* way. The reason for this is that many religious traditions around the world do *not* have a doctrinal focus on, or social norms about, belief in any sense. That is not to say that people in these traditions do not *have* beliefs in some sense or another. Rather, there is simply no doctrinal focus on the maintenance of one's mental state of "belief," which means that the people who practice these religions are unlikely to "believe in belief." To take a simple example, anthropologist Jonathan Lanman writes:

> My own experiences of problems with Western ideas of belief took place in Tokyo, Japan in 2018. There, one of the most common responses to being asked whether one believes in the existence of *kami* was "I don't know, I've never really thought about it," even by those who routinely visit shrines and temples.

Later, Lanman adds, "though many would, if pushed, assent to their existence."[47] (In the Shinto religion, kami are spirit deities that may include the spirits of some beings who were once people, like an emperor.)

Lanman's point is not that his informants *had* no beliefs about *kami*—evidently, they did—but rather that they were unaccustomed to having second-order views about whether or not they had beliefs on the matter in the way that Western Christians often do.[48] It follows, then—and this point generalizes to many non-Western religious traditions—that Lanman's informants who regularly venerated kami are unlikely to have anything like Dennett's posited attitude of belief in belief. Similar things could be said for Rita Astuti's informants among the Vezo in Madagascar, who *have* "belief" in some sense in their ancestors but appear unlikely to have the sort of doctrinal focus on having a certain mental state that could be labeled as "belief in belief." And so on.

None of this is to take away from Dennett's posit, which is likely a more or less accurate description of the mental state of many Western Christians. But even once we include it, there will still be many beliefs around the world that continue to pose The Puzzle without being amenable to Dennett's "belief in belief" notion that would in other cases alleviate it.

Solution 8: Weak belief. Another attitude solution would be to say that religious believers have only "weak belief" in their religious myths and doctrines. That would solve The Puzzle by saying that, although religious believers do (in some sense) factually believe, their level of belief is weak enough that the believers themselves can still be rational while maintaining it. As it happens, I know of no one who has explicitly maintained this as a solution to The Puzzle at hand. Yet some theorists have suggested it as an explanation for what I have called *compartmentalization* or *practical setting dependence*: the phenomenon discussed in Chapters 3 and 4 in which, outside the religious practical setting, religious people do not act as though their professed "beliefs" are true. Maarten Boudry and Jerry Coyne, attempting to criticize my earlier work, make the following suggestion:

> We do not hold that belief is a unitary phenomenon. Rather than a simple sort of cognitive attitude, "belief" may be a convenient label that lumps together a host of slightly different phenomena. First, belief varies in strength (i.e., degree of doxastic commitment). When it comes to religious belief, researchers have found a continuum.[49]

Perhaps, then, one could also appeal to lack of "strength" of belief to help resolve The Puzzle.

Relatedly, Jonathan Jong has published an article whose title is its main thesis: "Beliefs Are Object-Attribute Associations of Varying Strength."[50]

So the idea that people may weakly believe is attractive enough that it deserves consideration as a potential solution to The Puzzle.

On this potential solution, we would posit something like the following mental state:

Andrew <u>weakly believes</u> that the ancestors desire that he sacrifice an ox.

This, I think, may capture the mental state of some religious believers, and it does indeed solve The Puzzle *for those cases.* One can be a rational person and have a low, "weak" degree of belief in outlandish ideas, as long as there is a smidgen of evidence in their favor (e.g., people in the community often repeat them).

Yet this view also fails to solve The Puzzle in an entirely general way because it inevitably leaves one half of an important distinction hanging. Talk in the philosophy and psychology of religion about "strong" and "weak" belief often conflates two distinct mental features:

1. The degree to which a person is epistemically confident that a certain "belief" is true.
2. The degree to which that "belief" is central to one's identity.[51]

Crucially, many religious beliefs are strong in the second sense (identity), *without* being strong in the first (epistemic confidence). Recall Tanya Luhrmann's informant who once expressed her state of mind by saying something like, "Faith means I don't believe it, but I'm sticking with it." This person's "belief" appears to have been exceptionally weak in sense 1 but strong in sense 2. And this leaves proponents of the "weak belief" solution with a dilemma.

Either would-be proponents of the "weak belief" solution (i) dismiss the second sense of belief strength and weakness as an illusory construct or (ii) they don't dismiss it. If they go with (i), then they'll fail to make sense of a wide range of religious behaviors altogether. Not only are there the verbal reports of people like Luhrmann's informants; the nonverbal behaviors of many religious devotees reveal underlying religious "beliefs" that appear overwhelmingly strong in one sense—but weak in another. To return to Scott Atran's *Talking to the Enemy,* many of the Muslim terrorists he describes led dissolute lives: drinking, smoking, womanizing, selling drugs, and so on.[52] This makes their degree of belief appear weak. And yet positing a single layer of weak belief would fail to do justice to the fact that these actors perpetrated extremist violence *in the name of* their religious belief. So doing away with identity strength and weakness of belief deprives one of the ability to make sense of important facts. If, on the other hand, proponents of the "weak belief" solution go with (ii) and hence do countenance sense 2 (identity centrality), then they may say that even the extremists can have weak belief in sense 1 but strong belief in sense 2. But that would just amount to admitting that they have not offered a complete solution to The Puzzle, since they have done nothing to explain what sense 2 even amounts to. Furthermore, sense 2 apparently involves the positing of another attitude type altogether—which would effectively put them in my camp.

This brings us to the last solution, which I think will fill the theoretical gaps left by the other mooted solutions.

Solution 9: A distinct cognitive attitude. Is the person who pretends on-stage to be Hamlet doing something irrational? He clearly represents in his head something that is wildly false and contrary to all evidence: *I am Hamlet.* But he commits no irrationality in forming that representation. Similarly, if we return to the imaginative states of the children from The Playground parable in the Prologue, we can ask the same sort of question: Were they irrational to form their internal representations of Zalla, Hirgin, and Ghost? The answer is *no* because their internal imaginative representations were compartmentalized so that (both inferentially and behaviorally) they didn't impinge on the overall rationality of the children.

The distinct cognitive attitude solution to The Puzzle takes advantage of the fact that rational agents can still be rational while forming representations that are far out of keeping with their evidence, as long as they have an *attitude* that is distinct from factual belief toward the contents of those representations.

A natural thing to say, therefore, in solution to The Puzzle, is that religious people adopt an attitude that resembles imagining in terms of its evidential rationality constraints (i.e., weak to nonexistent) *and* in terms of its compartmentalization. To be *religious,* this cognitive attitude will have to have certain other properties that make it central to the religious person's identity. Most notably, it will guide symbolic action. But as we've seen in the previous chapters, there is no problem with the coherence of such a posit. And given the kind of creatures that we already know humans to be—rational, capable of secondary cognitive attitudes, and prone to forming group allegiances—it is plausible to posit such an attitude.[53]

Now let's compare this solution to the foregoing ones. Such a religious cognitive attitude can be taken toward outlandish contents, without the believer's being deluded. One also needn't be gullible to adopt this attitude. And it can be taken toward contents that are strange enough that they can't reasonably be defended by even the most ardent apologist. It thus fills the glaring gaps left by the solutions under the Adjust the Rationality Approach.

Such a religious cognitive attitude, furthermore, can be taken toward contents that are descriptive—the sort of "belief" contents for which Gould's NOMA solution was unable to account. And though this attitude defines one's group identity, one needn't reinterpret *all* of its contents as being *about* one's society, so this solution achieves the advantages of Durkheim's solution without the cost. A religious person can also take this attitude in relation to religious representations that are not so murky that they can't be evidentially evaluated, so it fills the gap left by Sperber's approach. And it can be taken toward *contents* at all, which allows us to avoid the desperate move of subscribing to noncognitivism. The distinct cognitive attitude solution thus also fills the theoretical gaps left by the Adjust the Content Approach.

Finally, such an attitude can be taken by religious persons who do not have an explicitly thought-out doctrinal focus on having beliefs, so it fills the theoretical gap left by Dennett's "belief in belief." And it is "strong" in the sense of centrality to one's identity, even if one's epistemic confidence concerning its contents is relatively weak—which, as we saw in Chapter 4, often happens.

Thus, the posit of a distinct religious cognitive attitude type—religious credence—is a crucial piece in solving The Puzzle of Religious Rationality. It is a posit that can be independently defended, as I have done, and it fills every gap left by the other solutions under consideration. The Puzzle of Religious Rationality is thus a strong further motivation for accepting that there is such a cognitive attitude as religious credence, which is distinct from factual belief.

Epilogue

The Playground Expanded

Though fictional, the parable of the Playground, set out in the Prologue, involves a lighter editing of reality than you might have thought.

When I was in first grade, there was a small group of kids who played with G.I. Joe toys in the shaded sand under the newly built, large wooden play structure on the playground at Oakdale Christian School in Grand Rapids, Michigan. To me, this group seemed like the cool kids. But I hesitated to try to join for some time because I had none of the elaborate G.I. Joe toys they did—tanks, fighter planes, and so on. But one day, I did acquire something that made me just confident enough to try to join the group: a Zartan figure. Zartan figures, though they were neither large nor elaborate, were new at the time and comparatively rare; no one in the group of kids had one yet. So when I approached, despite my trepidation, they welcomed me. "Cool, Neil has Zartan!"

But the next day, another kid tried to join the group. Timidly, he approached, trying to work his He-Man figure into the playing we were doing with our G.I. Joes. The boy was not antecedently less popular or likable than I had been the day before. But his bid to play with us was swiftly rejected. "We don't want He-Man!" said the leader of our group. The shady space beneath the wooden play structure, in our eyes, was *ours*, where who *we* were was defined by which game of make-believe we were able to play and serious about playing. No one in our group protested the banishing of the boy with the He-Man toy. He left, pretending not to be hurt.

The idea of this book—and it still astonishes me—is that considerable fragments of the multifarious social phenomena we call religions are large-scale versions of *that*: games of make-believe that define a group.

Consider how I *felt* when the G.I. Joe group let me join: I felt accepted, I felt like I belonged, I felt like I was not alone, I felt like I had a home. We might even call these feelings *religious*. The content of the play itself was fun. After all, we were enacting epic struggles between competing powers. But that which we imagined and enacted—those superheroic agents, actions, and events—was, at a deeper psychological level, a way for us to be together. And so it is, much of the time, with religion. That idea is the Durkheimian core of this book.

But unlike Durkheim, I see no need to reinterpret the apparent *contents* of religious "beliefs" as being about society. Our imaginings of G.I. Joe were *of G.I. Joe* and didn't have *us* as their contents; likewise, religious credences do characteristically represent what they appear to represent: gods, spirits, ancestors, floods, reincarnations, paradise, hell. It is rather our attitude toward these contents—religious credence—that enables them to guide symbolic action in concert with sacred values in ways that cement group cohesion. Yet that attitude is still compartmentalized enough that it doesn't typically lead to rampant confusion about the empirical and the humdrum, with which, as Evans-Pritchard puts it, people are concerned nine-tenths of the time.

*   *   *

I am sure I have not convinced all my readers. I have no doubt I will be hearing quarrels and objections for as long as I continue to discuss these topics, which, given my proclivities, probably means the rest of my life. Furthermore, I find it entirely likely that I got some details wrong—both of theory and of fact. Mistakes at various levels are nearly inevitable in work such as this that crosses so many fields, each with its own high level of expertise.

Yet despite my awareness of opposition and of the likelihood of flaws, I am confident that my overall picture is correct. The reasons for this confidence can be encapsulated in four points. One: we know that humans are creatures who play pretend from a young age. Two: we know humans are creatures with florid imaginations that drive such make-believe play—along with many different forms of pretending well into adulthood. Three: we know humans are creatures who form packs, clans, tribes, teams, groups, and . . . religions. And four: we know that humans create and adhere to symbolic representations that define the boundaries of such coalitions—symbolic representations that range from ritual actions in the world, to doctrines and stories that are spoken or on the page, to mantras that are internally recited. I said above that I still find my view astonishing. But that is only from the perspective of the prevailing philosophical currents against which it swims. From the standpoint of the four points listed in this paragraph, it would be astonishing if we humans *didn't* use our capacity for imagining to form sacralized secondary cognitive attitudes—sacred imaginings—distinct from factual belief.

In other words, there is every reason to be confident that religious credence, in a form at least something like I've characterized it, exists. The question is how widespread it is. I have used psychological, anthropological, and historical data to argue that it can be found across cultures and well into the past, both in organized religions and in small-scale societies, and that many "beliefs" in the heads of religious practitioners are in fact just such religious credences. This does not dimmish them or make them any less important; it just says more clearly what they are.

Furthermore, because religious credence is an attitude, it can, and almost certainly does, occur in relation contents that are not overtly "religious," as I made plain in already Chapter 1. Just as the Azande may have had religious credences concerning the deliverances of their poisoned chicken oracle, so, too, may many recent Americans have had (and still may have) religious credences concerning the deliverances of Q. Many, of course, may have had genuine factual beliefs in either situation. But once we are clear on what religious credence is and that it is a compelling psychological attractor position, there is every reason to appeal to it in formulating hypotheses about the underlying nature of the odd and oscillating "beliefs" that crop up in the social world around us. That, more than anything, is what I hope my contribution to research amounts to: a better conceptual tool kit for crafting hypotheses about "belief" than we had before.

*   *   *

There are a few broad lessons I haven't stated yet and want to state now. We are looking toward the horizon; having crossed the territory we crossed, we are now in a position to make out some shapes that we couldn't see at all before.

First, the psychological work of this book opens up serious questions in epistemology that until now couldn't have been asked directly and clearly. Epistemology over the last fifty years has been largely preoccupied with questions like: *What makes beliefs justified? What makes beliefs rational? What are the norms for beliefs? Under what conditions can beliefs be counted as knowledge?* And so on—the questions are about what makes "belief" epistemically good or at least better in some way. But if my basic position is right, most of these questions are either ill-formed or have an enormous blind spot. If, on the one hand, one interprets the word "belief" in these questions as referring to *every* mental state that can be pretheoretically called "belief," then the questions are ill-formed. The reason is that, interpreted this way, they lump together vastly different mental states: religious credence versus factual belief (and perhaps other quite different states as well). But those cognitive attitudes are so different that we have no right to presuppose that what justifies the one is *the same sort of thing* as justifies the other, contrary to what the epistemological questions (on this interpretation) tacitly presuppose. Is a religious credence that *p* to be justified in the same way as a factual belief that *p*? I have no idea, but at least I recognize it *is* a question. (And, recall from Chapter 5 that laypeople appear *not* to presume that religious and factual beliefs have the same norms.) If, on the other hand, one takes the questions in a narrower way, as referring to the sort of thing I call factual belief, then the questions have an enormous blind spot: we can answer them thoroughly without having the slightest idea of what norms aptly apply to religious credence. There is some reason for thinking the questions are often (tacitly) understood in this narrower fashion: the diet of examples in analytic epistemology tends to include beliefs with dreary contents like *that I am looking at a barn, that the chair is blue,* and (famously) *that the man who will get the job has ten coins in his pocket.* These sorts of contents are far likelier to go with a matter-of-fact belief attitude, so it is reasonable to say that that was the target all along, in which case what epistemology has done these past fifty years is thoroughly explore justification (and other

good-making properties) for *one type of* descriptive belief attitude. That is
no small thing. But again, is a religious credence that *p* to be justified in
the same way as a factual belief that *p*? We don't know. Related questions
arise. For example, if we grant the reasonable and standard view that factual beliefs can partly constitute knowledge (when they are true, justified,
and whatever else is needed to clear the bar), and if we also grant that religious credence is not factual belief, we are left with the following striking
question: can religious credences also partly constitute knowledge in the
same way? The answer is just not obvious. Many would doubt that a justified, true, <u>imagining</u> (which is certainly a possible state to be in) amounts
to knowledge; the difference in attitude seems relevant to whether the
mental state as a whole can be knowledge. So, what to say about religious
credence? At least now we can ask the question clearly. The greater lesson
is that the exercise I just went through in this paragraph can be done for
almost *any* big normative question about "belief" in epistemology. It's like
turning one's head and realizing the territory to be explored is far bigger
than one thought.[1]

Second, the work of this book enables us to characterize a distinctive
form of self-deception that can, and I think does, arise in religious and
ideological contexts. I wrote a great deal about self-deception at the start of
my career, but I stopped because I found that the term was slippery enough
to invite cross talk (even more so than "belief" does).[2] Furthermore, talking of self-deception can usually be replaced with talking directly of the
various biases that arguably compose it. Nevertheless, I think it is useful
for present purposes to retool my earlier notion. The self-deception I wish
to characterize is this: a person who has a religious credence that *p* may,
due to various motivations and social pressures surrounding people's "beliefs," try to convince themselves that they factually believe that *p*. Such
motivations and pressures likely do not exist in all cultural contexts. But
they exist in many. Consider this passage from 1 Corinthians, which appears in the communion liturgy of the Christian Reformed Church—the
denomination to which my family belonged most of my life and in which
my father is still ordained:

> Whoever eats the bread or drinks the cup of the Lord in an unworthy
> manner will be guilty of sinning against the body and blood of the Lord.
> Everyone ought to examine themselves before they eat of the bread and

drink from the cup. For those who eat and drink without discerning the body of Christ eat and drink judgment on themselves. [1 Corinthians 11:26–30, NIV]

The phrase "unworthy manner" was taken to mean something like *lacking belief,* and the word "judgment" raised the specter of *damnation.* Now recall from Chapter 5 that people generally have the ability to think about nuanced differences in cognitive attitudes and differentiate different kinds of "belief." And, recall from Chapters 3 and 4 that religious "belief" commonly coexists with *doubt* about what is believed. Against this background, it is very easy for Christians who hear such verses in their communion liturgies to be plagued by the following fearful thought: *maybe the kind of belief I have isn't enough.*[3] That is hard to live with, so against that background, it would not be surprising if many Christians self-deceptively tried to convince themselves that they have a kind of belief about, say, reincarnation or seven-day creation that they actually don't; namely, factual belief. Such trying is by its nature self-deceptive.[4]

I have focused on Christians in illustrating the form of self-deception that this book puts us in a position to characterize—that of trying to conceive of one's religious credences as though they were factual beliefs—but that is only because the example is stark and intimately familiar to me. But we live in an age when this type of self-deception is likely to be all around us. The intensity of political and ideological partisanship in the contemporary United States, especially, puts nearly everyone in the position of feeling a great deal of pressure to have certain specific "beliefs" on more and more issues: gender, race, vaccines, abortion, policing, income inequality, American history, election fraud, standardized testing, guns, and even (for the love of God) the efficacy of surgical masks (see: didn't I tell you *anything* could be sacralized?). So on any socially divisive issue and for whatever team you're on, you are likely, with respect to at least some of the "beliefs" that your team requires of you, to find yourself with the same fearful thought: *maybe the kind of belief I have isn't enough.* And then you feel guilty. If you find yourself having such fearful thoughts with respect to any doctrine or ideological issue, your position is entirely understandable. Yet know that you're sliding toward self-deception. And there is a risk then, realized commonly enough, that one will start bullying others about their "beliefs" in service of deceiving oneself. That is ethical harm in service of epistemic irrationality. I don't commonly take strong normative stances—I mainly try just to describe what is going on clearly—but I think it's fair to say that any religion or ideology that makes you feel pressure to do *that* is pernicious.

So what should you do if you find yourself, for any religious or ideological proposition to which you feel pressure to subscribe, with that very fearful thought (*maybe the kind of belief I have isn't enough*)? I think you should ask yourself: *enough for what?* And if the thing you're going for is *worth* going for, it should be attainable without self-deception. Figure out how. When I do this exercise (asking *enough for what?*), I find my anxiety about my own doxastic state is transformed—*poof!*—into curiosity. I hope it does the same for you.

Third, it is not only religious people (or proponents of an ideology) *themselves* who are motivated to cast what are in fact religious credences as factual beliefs. We often see the mirror of this in *opponents* of religion. There is some heavy irony here. Consider this passage from Sam Harris's *Letter to a Christian Nation.*

> According to a recent Gallup poll, 12 percent of Americans believe that life on earth has evolved through a natural process, without the interference of a deity. Thirty-one percent believe that evolution has been "guided by God." If our worldview were put to a vote, notions of "intelligent design" would defeat the science of biology by nearly three to one. This is troubling, as nature offers no compelling evidence for an intelligent designer and countless examples of *un*intelligent design. But the current controversy of "intelligent design" should not blind us to the true scope of our religious bewilderment at the dawn of the twenty-first century. The same Gallup poll revealed that 53 percent of Americans are actually *creationists.* This means that despite a full century of scientific insights attesting to the antiquity of life and the greater antiquity of the earth, more than half of our neighbors believe that the entire cosmos was created six thousand years ago. This is, incidentally, about a thousand years after the Sumerians invented glue. Those with the power to elect our president and congressmen—and many who themselves get elected—believe that dinosaurs lived two by two upon Noah's ark, that light from distant galaxies was created en route to the earth, and that the first members of our species were fashioned out of dirt and divine breath, in a garden with a talking snake, by the hand of an invisible God.[5]

As I see it, Harris is portraying people's religious credences as if they functioned psychologically like factual beliefs—so that he might better vilify their craziness. In fairness to him, many American Christians (the ostensible audience of his *Letter*) have a lot to answer for. And it may also be that their worldview has *prevented* a good many of them from learning things about science that they otherwise would have learned. But as I made clear in the last chapter, it is an error to impute the sort of lunacy implied in this passage to millions of people (*53 percent of Americans think there was a talking snake!*) on the basis of what people profess verbally.

In this conflation of religious credence and factual belief, strident atheists and fervent religious apologists are strange bedfellows, both taken in by the cause (though on opposite sides of it) of "defending" or "refuting" religious ideas as would make sense *if* the attitude that the faithful invariantly took toward them were factual belief. So the strident atheists and religious apologists keep each other in business through a shared ignoring of a psychological difference of which the majority of laypeople, as I showed in Chapter 5, are at least vaguely aware. In the end, I see nothing wrong with taking issue with or trying to defend religious ideas. But if one wishes to do that, the better part of honesty is to be clear about what *attitude* one's interlocutors have in relation to those ideas—and what attitude one has oneself.

That brings us to the fourth and final lesson. I hope that the work I've done in this book is a tool not just for crafting hypotheses in the course of research but also for self-knowledge. That, after all, is both good in itself—at least for those of us who give credence to the Delphic Oracle's main injunction to *know thyself*—and the antidote to the forms of self-deception just described. In addition to being inherently in conflict with self-knowledge, there is a further interesting reason why self-deception may be bad in itself. Stephen Darwall distinguishes three approaches to moral theory: approaches based on conceptions of what is *good,* approaches based on *duty,* and approaches that have as their foundation the *constitution* of the moral agent. The constitutionalist approach, forcefully represented by Kant, holds that morally right action is that which comes from an agent with the right constitution—one that is set up to appraise actions fairly and impartially. If there is anything to this, then we have serious moral reason to strive against self-deception. Darwall writes that self-deception, on the third moral approach, "is not only a misuse of judgment, it threatens the very capacity for judgment. Constitutionalists must regard it, therefore, as both wrong in itself and a threat to the very possibility of moral integrity."[6] That wording is weighty, but I think the point can be captured, perhaps imprecisely, in the idea that one of the most morally important things people can do is not be alienated from ourselves—which includes not deceiving ourselves about what our "beliefs" are really like. In that spirit, I hope that the ideas laid out in this book occasioned a movement toward a better understanding not only of the minds of others—but also your own.

Notes

PROLOGUE: THE PARABLE OF THE PLAYGROUND

1. Why say that there are *two* maps as opposed to one larger, complex one? If there were just one cognitive map, it would be riddled with confusing inconsistencies. Are the doll characters hand-sized or larger than life? Are they immobile, or do they propel themselves? Are they mute, or do they have loud voices? If all these representations were part of one big cognitive map of the situation at hand, that map would be utterly baffling. But given that there are two distinct cognitive maps in the mind of each pretender, the pretender can use one map in guiding one fragment of pretense activity and the other in guiding other portions of play. A crucial skill in pretend play is to wink at the apparent inconsistencies between the two maps while still employing both of them in guiding one's pretense actions.

2. See Leslie (1987) on how even children avoid "representational abuse." Gendler (2003, 2006) also recognizes that "imaginative contagion" is the exception and not the rule. And Liao & Doggett (2014) compellingly critique Schellenberg (2013) for suggesting that imagining and belief get blurred in cases of pretense immersion.

3. Golomb & Kuersten (1996).

4. M. Taylor (1999: 14–15).

5. M. Taylor (1999: 116).

6. People who say that some fictional entities are "real" to someone are often making the point that that person doesn't have full voluntary control over the way ideas about such entities unfold; the imaginary takes on a life of its own, so to speak. This is a fine point that is misleadingly expressed, however, since it errone-

ously seems to imply confusion between fantasy and reality. A better way to put it would be that the imaginary can "feel real" as opposed to the ambiguous phrase "being real *to* so-and-so." Another genuine phenomenon that might be captured by phrases like "real to her" is *absorption* (cf. Luhrmann, Nusbaum, & Thisted, 2010). But absorption also does not entail confusion since people often get absorbed and immersed in representations they know to be fictional—the two maps stay distinct throughout the immersed play—as Liao & Doggett (2014) argue convincingly.

7. The points in this paragraph rest on a large body of insights about the relation between rationality and belief (understood in the first-map sense, in my terms) that come from Davidson (1984), Dennett (1987), Dretske (1983), and Williams (1973)—among others.

8. Plato makes an interestingly related point in the *Phaedrus*, when he has Socrates argue that in order to lie well, a person must know the truth. Of course, if one views lying as a form of pretending, that just becomes a special case of the point I made in the paragraph that contains this endnote.

9. As Walton (1990) famously argues.

10. P. L. Harris (2000: ix); my italics.

11. See Van Leeuwen (2009a) for more on this point.

12. This description of a "vision" along with the small group discussions about it that follow in starting the new "religion" is based loosely on the accounts of founding events that appear in Ann Taves's (2016) book *Revelatory Events*. Note also that the collaborative creation of an imagined world that I portray here also resembles the generation of paracosms among friends in middle childhood that Marjorie Taylor describes in her more recent work. Describing the creative work of two boys on a shared paracosm called "Abixia," she and her colleagues write: "M1 has minted Abixian coins, printed Abixian money, created a dictionary of the Abixian language and written a gospel for the Abixian religion. M2 has made sculptures and architectural drawings" (Taylor et al., 2015: 169).

13. Taves (2016) portrays how the subsequent, official, socially conditioned interpretations of founders' visions often depart from what those initial visions must have been like; she draws out that contrast in the cases of the founding visions of Mormonism, Alcoholics Anonymous, and A Course in Miracles.

14. I put "heard" in scare quotes because often the "hearing" of a divine voice is more like auditory imagery. See Luhrmann's "But Are They Crazy?" (chapter 8 in Luhrmann [2012]); see also Taves's (2016) discussions of "The Voice" that inspired A Course in Miracles.

15. A brief note on plausibility: The Playground includes many psychological elements of religions around the world, as studied empirically. On minimally counterintuitive deities, see Atran & Norenzayan, 2004; Norenzayan & Atran, 2004; Boyer, 2001: chap. 2; McCauley, 2011: chap. 4. On relevance and access to human affairs, see Boyer, 2001: chap. 4. On taking a moral interest in human affairs

(which many but not all deities do), see Norenzayan, 2013. On gods as agents (often anthropomorphic), see Guthrie, 1993; Barrett & Lanman, 2008; Van Leeuwen & van Elk, 2019 (of course, the observation about anthropomorphism goes all the way back to Xenophanes).

16. As should be obvious, I am using the word "credence" here as a term of art for a quite different construct from the 0 to 1 degree of belief construct that appears in formal epistemology, which theorists in that context also often call "credence." Throughout this text, I will only use the word "credence" in my specific fashion and not in the other way, so there is no risk of confusion.

17. There is also a question of the extent to which people are metacognitively aware that their religious credences differ from factual beliefs. I suggested at the outset that people have some level of intuitive awareness, and I handle that question more extensively in Chapter 3, Chapter 5, and the Epilogue. To preview: there is ample evidence that people in a variety of cultures *do* regard their religious "beliefs" as involving a different attitude from their factual beliefs, though—as I discuss in the Epilogue—in some religious traditions, there may be cultural pressure to avoid recognizing that difference.

18. A note for replication afficionados: I and my research assistants vetted all the empirical studies incorporated into the argument of this book to make sure *none* of them have replication attempts that failed. At the same time, I included many studies that (at the time of writing) had yet to be replicated, simply because they were highly pertinent and the best thing one had to go on for a particular topic. My aim was thus to avoid both excess caution (which would exclude much valuable information) and to avoid excess trust (which would include spurious results) when it came to taking empirical results seriously. Research, after all, is an ongoing project.

19. Cf. Kuhn (1962: chaps. 3–4).

20. Boyer (2001: 6–10).

1. THE ATTITUDE DIMENSION

1. Since this book is not mainly about content, I'm happy to settle for a fairly rough-and-ready characterization of that notion. For a more developed view about how content should be related to psychological theorizing, see Egan (2014), which I am happy to endorse.

2. Of course, I'm not implying that doubting that p and hoping that p are mutually exclusive; they *often* co-occur. The point is that they are different attitudes.

3. Goldbach's conjecture is that any even number is the sum of two primes. At the time of writing, it has not been proven, nor has a counterexample to it ever been found.

4. Of course, attitudes in the social psychological sense will often be *in part composed of* many attitudes in the sense I am developing.

5. This example is not merely hypothetical. Ingela Visuri, who does qualitative research on religious experiences among autistic informants, points out that one of her informants up into his teenage years actually thought unicorns existed (Coleman & Visuri, 2018).

6. If suspension of judgment is an attitude, it may be an exception to this characterization of cognitive attitudes, but not one that will make a big difference to the overall picture I present here (cf. Friedman, 2013).

7. This way of describing the difference between cognitive and conative attitudes is inspired by—but has some differences from—that found in Shah & Velleman (2005).

8. Another example, mentioned in both Sperber (1997) and Van Leeuwen (2014a), is "jade." The word "jade" refers to two quite different chemical substances that have surface similarities (jadeite and nephrite, in technical terms) and thus are not clearly distinguished by pretheoretic language.

9. The locus classicus of this framework is Fodor (1985) (see Schwitzgebel [2001, 2002] for the contrasting dispositionalist view). I have adjusted Fodor's notation slightly for my own ends. Fodor uses "aRmp" instead of "aRmc" for his abbreviation, where "p" is for *proposition*. I prefer "c" for "content" because many religious "beliefs" have a different sort of content (many philosophers might even be hesitant to call it "content" at all), which Sperber (1982) calls *semipropositional*, where the "belief" (given its verbal expression) appears to have *something like* propositional content, but it is hard to assign one exactly (e.g., when one believes something that is expressed as "God is three persons in one," it's not exactly determinate what the believed proposition even is, so the belief's content is a semiproposition). This issue will become relevant in Chapter 8.

10. True, for most cognitively well-functioning individuals, their broad background of factual beliefs is largely accurate (on pain of their not being successful in life), but that is a *substantive* fact and not a *definitional* one.

11. One might wish to call Kai's mental state an *ideological credence* rather than religious credence, which is fine. Just keep in mind that those are two different ways of talking about much the same attitude type—what matters for our purposes is the distinctive manner of processing rather than its label.

12. See Sperber (1996) and McCauley (2011) on attractor positions; see also McCauley & Lawson (2002).

13. Porot & Mandelbaum (2021: 12). Boudry & Coyne (2016a) make the mistake as well. See Van Leeuwen (2018) for discussion.

14. See Van Leeuwen (2014a, 2017b).

15. Bloom (2015) floats this suggestion in a popular *Atlantic* piece.

16. Mandelbaum (2014, 2019).

17. See Ranney & Clark (2016) and Weisberg et al. (2021).

18. In fact, as you will see, it is part of how I *define* factual belief as an attitude concept that factual beliefs respond to evidence, so the real question is whether

mental states with those properties (responsiveness to evidence and related ones that I use to define the notion in question) do exist (I argue that they do in Chapter 3). As is usual in psychological theories like mine, there is a definitional component and an empirical component (see Lewis [1983: 111] for a canonical statement of this method): the definitional component involves using a theory to posit certain phenomena, which are labeled with appropriate terms, and the empirical component involves gathering evidence that phenomena in the real world satisfy the theory. So not only have Porot and Mandelbaum illegitimately inferred attitude type from content type (though those are independent dimensions); they have also conflated a pretheoretic use of the phrase "factual belief" with my technical use of that phrase as a term with a specific meaning in my theory.

2. A THEORY OF COGNITIVE ATTITUDES

1. Hume, *Enquiry,* V, Part 2.

2. Here's another example of a characterization of belief that presupposes what it ought to explain. Schellenberg (2013) characterizes believing that p in the following terms: belief is "a mental state of taking to be true" (499). But what is the significance of the word "taking" here? What does "take" even mean? Of course, that word is general purpose, and she doesn't specify how to understand it. So, for that formulation to sound right, the reader has to supply the relevant concept for interpreting the word "take." And the concept to be supplied must be belief in some sense, for none other is given and none other would do. Schellenberg's formula doesn't so much explain that concept as tacitly invoke it, as I explain in Van Leeuwen (2014b).

3. See Van Leeuwen (2013) for a discussion of how the word "imagine" has different senses. It can have an *attitude* sense, a *constructive* sense, and/or an *imagistic* sense. These are orthogonal. The sense in question here is the attitude sense; that is, "imagine" means (roughly) to entertain an idea in a nonbelieved way (where what that amounts to more exactly is what this chapter aims to spell out).

4. Consider Gervais & Norenzayan (2012), who have participants rate their religious beliefs on a 100-point scale. It's not clear what that scale even means (as it happens, their well-known experiments also failed to replicate).

5. Clifford (1877/1999).

6. For a defense of the idea that flat-out belief and degrees of belief play distinct and separable roles, see Buchak (2014) and Ross & Schroeder (2014). For discussion, see Jackson (2020).

7. Davidson (1963). How does this action-theoretic view connect to Hume's other view that belief is a more forceful and vivacious sentiment? On my reading, the sentiment Hume posits is supposed to *explain* belief's action-theoretic properties. My assessment of the two views in question is that Hume is wrong about

the forceful and vivacious sentiment, but he is on the right track about the action theory, but more specification of belief's causal functional role is needed.

8. This is my example, not Velleman's, but his examples point in the same direction.

9. See Velleman (2000).

10. I develop this issue in Van Leeuwen (2009a). Also, Shah & Velleman (2005) is a valuable resource for understanding why it's difficult to distinguish belief from other cognitive attitudes. See also Bratman (1992).

11. Sperber (1982: 171).

12. B. Williams (1973) is the locus classicus for involuntarism about belief. See Levy & Mandelbaum (2014) for a more recent discussion of the voluntarism issue.

13. As I argue in Van Leeuwen (2009a).

14. Gendler (2003, 2006) discusses various forms of "contagion," or ways imaginings influence behavior in the manner beliefs typically would. Importantly, however, she thinks this is the exception to a general rule (for otherwise, the distinction between the attitudes would collapse).

15. The theory-construction strategy here is implicit definition. I'm defining attitudes by way of asymmetric and antisymmetric relations between them, as David Hilbert did in geometry with *point, line,* and *plane.* The relations allow us to single out the attitude types of interest even though they are defined in terms of each other.

16. For an excellent example of a closely related strategy, see Dub (2017).

17. Sperber (1996); McCauley (2011); McCauley & Lawson (2002). See also Van Leeuwen & Lombrozo (2023) for a statement of the relevant methodology.

18. See Ginet (2001), Frankish (2004), and Weatherson (2008).

19. Quattrone & Tversky (1984).

20. Batson et al. (2007).

21. Though interestingly, there still was deception in this experiment. The participants were deceived about the existence of the other person who would be receiving shocks. Evidently, the experimental paradigm required them to believe in the existence of the person whose (potential) shocks they were to imagine.

22. See Mele (2001) and Van Leeuwen (2008).

23. Bratman (1992).

24. Harris (2000: 12).

25. See Lillard & Witherington (2004).

26. See endnote 15 for an explanation of why it is legitimate to invoke factual belief in this stage of theory construction, even though, from a broader perspective, that is one of the attitudes we're trying to characterize.

27. Walton (1990).

28. Harris (2000: 13); my italics.

29. This picture parallels D. Lewis's (1973) theory of counterfactuals, where factual beliefs parallel the actual world and imaginings parallel nonactual possible

worlds that are more or less close in similarity to the actual. On this picture, a relatively realistic imagining is the psychological analogue of the truth conditions of a relatively realistic counterfactual.

30. See Kind (2016) for a paper that comports with this view, along with Van Leeuwen (2016a).

31. See Van Leeuwen (2013).

32. See Currie & Ravenscroft (2002: 18–19) for a clear statement of this view.

33. Goodstein & Weisberg (2009); Weisberg et al. (2013).

34. Stock (2017: chap. 6) is an example of someone who questions my view that beliefs govern imaginings in the way I identify here.

35. One could, of course, imagine some avant-garde fiction in which all of those factual beliefs were violated continuously, but my point is that the further down that road one goes, the less the sentences of this fiction (and hence the less the imaginings about the fictional world) have content at all.

36. It's technically *antisymmetric* there rather than *asymmetric* since the relation I'm talking about is reflexive (factual beliefs supply information to other factual beliefs as well) and antisymmetric relations allow for reflexivity, while asymmetric ones (by definition) technically don't. People not immersed in mathematical logic can ignore this wrinkle without a loss of understanding.

37. There are three obvious ways that imaginings influence factual beliefs, but none of them is cognitive governance in the sense I specify. First, one often forms factual beliefs *that* one is imagining something—that is, *about* one's own imaginings. This is obviously not cognitive governance in the defined sense any more than my forming beliefs about *your* imaginings implies your imaginings cognitively govern my beliefs since the belief contents that emerge are second order. Second, imagining is often a way of coming to realize things one otherwise might not have. If I were a detective, I might imagine that a murder was done with poison and thereby come to realize (hence believe) that it was done with poison. But I would only draw this conclusion if the imagined manner of crime cohered with other beliefs I have about the facts of the situation. I don't simply import the imagining's content for the sake of drawing inferences among beliefs; rather, I consider those contents, and the imagining leads to a hypothesis that is confirmed by other beliefs. This is also not a case of imaginings cognitively governing beliefs; it is really a case of beliefs cognitively governing themselves, with imagining being used in an exploratory, not governing, fashion. Third, there are false memories that originate with imagining something, but that is a performance error and not inference, so it is not cognitive governance.

38. I do this because any individual mental state in a class may be bracketed for purposes of inference, even when a class as a whole provides the default source of information for guiding inferences.

39. D. Lewis (1978) and Walton (1990) both have elaborate discussions of how fictional "truths" are generated, and parallel complications arise in the generation

of imaginings. Nonetheless, it remains true that factual beliefs are bedrock. So, for example, even when genre truths take the place of (some) factual beliefs in supplying background information for purposes of inference about what's "true" in a fiction, it is still the case that one has <u>factual beliefs</u> to the effect that those are the genre truths, and so on.

40. Dennett (1987); Davidson (1984); Dretske (1983).

41. Stich (1981) cites, for example, Nisbett & Ross (1980) and Wason & Johnson-Laird (1972) as important results in social psychology that document human irrationalities.

42. Stich (1981); Johnston (1988); Bortolotti (2005, 2015); Mandelbaum & Quilty-Dunn (2015).

43. Gendler (2007); D'Cruz (2015: 980).

44. As I argue in Van Leeuwen (2018) "The Factual Belief Fallacy."

45. We should also note that arguments from self-deception and other forms of irrationality, like those in Johnston (1988), don't establish their desired conclusion. Consider self-deception. A self-deceived businesswoman might believe profits are just around the corner, even though a neutral view of the evidence suggests no such thing. So how does she accomplish this self-deception? As I've argued elsewhere (Van Leeuwen, 2008), she attends to the teaspoon of evidence that supports her self-deceptive belief (a few good days of sales) and ignores the mountain of evidence to the contrary (many bad days of sales). But that process would *only* make sense if beliefs—even self-deceptive ones—were tethered to evidence in a rational way to some degree.

46. Imaginings, of course, have some constraint, as Kind (2016) argues. But they are far less constrained than beliefs, and their constraints are often defaults that can be voluntarily, if effortfully, altered.

47. I acknowledge that there is a slight redundancy in my theory insofar as I claim involuntariness explicitly and have it in my subclauses in Evidential Vulnerability 1. I do this for two reasons. First, starting with involuntariness helps readers focus on the class of beliefs of interest; second, the arguments for the properties are usefully separated.

48. Gopnik (2003: 240).

49. See Povinelli (2000: chap. 3) for the comparison. See de Waal (2016: chap. 2) for counterpoint.

50. Baillargeon (2002: 54–55).

51. Following Spelke (1994) and Leslie (1994, 1995).

52. Harris summarizes the research extensively in Harris (2012) *Trusting What You're Told*.

53. Kim, Kalish, & Harris (2012).

54. Cf. Gelman, 2003; Gelman & Coley, 1990.

55. An anonymous reviewer for Harvard University Press pointed out in response to an earlier version of this manuscript that in some instances, young children's beliefs (seemingly factual beliefs) can resist updating in light of negative feedback/evidence. For example, they will expect a ball dropped into a tube that is sloped to come out directly below where it entered the tube rather than at the actual exit of the tube, and this "gravity error" can persist for some time, despite repeated exposure (Hood, 1995). I respond to this point as follows: for a factual belief to fulfill its role of updating in light of evidence, the person who has it must have the *capacity* to cognize a given event or piece of information *as* relevant evidence; such capacities do not always come online right away, which is part of why we often see the developmental lags we do. So the relevant attitudes may still be vulnerable to evidence in my defined sense, but we won't see proper updating until a child has the relevant cognitive capacity. And, indeed, people do *eventually* learn that the ball will follow the slope of the tube, which exhibits the evidential vulnerability of the relevant factual beliefs after all.

56. Armstrong (1973) was a champion of this slogan, inspired by Frank Ramsey (1929/1931). But it is also common in everyday talk about belief among contemporary philosophers.

57. These basic properties of factual beliefs hang together in systematic ways. For example: evidential vulnerability goes with practical setting independence because we need to update our basic map of the world for the sake of avoiding ditches and trapdoors *no matter what setting we're in* (even or perhaps especially if we're pretending we're in a place without ditches or trapdoors). Evidential vulnerability also goes with involuntariness since being tethered to evidence is the constraint on factual beliefs that the voluntary, secondary cognitive attitudes don't have. Evidential vulnerability also goes with cognitive governance since, for anything we want to imagine, to imagine it well, we should have some evidence-responsive ideas about how that thing really is. Otherwise put, the psychological features that constitute factual belief form a natural kind in virtue of being a *homeostatic property cluster*—a cluster of properties that, though logically independent of one another, typically co-occur for systematic reasons that have to do with the causal structure of the world (Boyd, 1991).

58. Let me briefly highlight a methodological point here. It would be easy enough to craft a loose theory of "belief" that papers over the distinction that I draw and thus lumps all "beliefs" together. Would that undermine my distinction? No, because it could only do so by shirking systematic philosophical obligations. Show me a theory of cognitive attitudes that actually satisfies Hume's Desideratum and thus distinguishes factual belief from *imagining,* and I'll show you a theory that, in light of empirical evidence, places many religious "beliefs" on the imaginative side of the divide.

3. RELIGIOUS CREDENCE IS NOT FACTUAL BELIEF

1. Boyer (2001: 86); my italics.

2. His more recent position seems to have shifted to something closer to a Two-Map View; see Boyer (2013).

3. Boudry & Coyne (2016a: 602) and (2016a: 613), respectively; their italics. See also Boudry & Coyne (2016b).

4. Levy (2017, 2020) also defends a One-Map Theory, which seems to be quite general. S. Harris et al. (2009) defends a One-Map View; it's not clear how general they intend it to be, but Harris's popular writings suggest quite general. Luhrmann (2018) discusses some One-Map Views in anthropology, with which she disagrees.

5. I should grant that even the most general views in these areas will admit some exceptions if for no other reason than to allow for people who are atypical. Still, the thrust of Boudry and Coyne's view is clear in trying to be as general as possible.

6. Note, however, that the claim can still be developed in different ways. Levy (2017) holds a general One-Map Theory according to which features of mental state *content* (or representational structure) entirely explain why religious beliefs don't function like straightforward factual beliefs do: since ideas of the supernatural are "unintuitive" and subject to "disfluent" processing, they don't feed into behavior as straightforwardly as the factual belief, say, *that there is milk in the fridge*. I respond to this in Van Leeuwen (2017a).

7. Luhrmann (2012); Bialecki (2017).

8. Bialecki (2017: 34); my italics. Luhrmann (2012: 100) anticipates Bialecki's emphasis on double coding by writing about how Vineyard members have an "epistemological double register."

9. Goffman (1959/2002: 207); my italics.

10. Bialecki (2017: 147).

11. I thank one anonymous reviewer for HUP for raising this concern.

12. See Luhrmann (2012: 62) for an interesting development of this example.

13. Luhrmann (2012: 65).

14. Luhrmann (2012: 70); my italics.

15. Luhrmann (2012: 70).

16. Luhrmann (2012: 131).

17. For the development of the relevant notion of props, see Walton (1990).

18. Levy & Mandelbaum (2014: 16). Note that by "believe," they mean more or less what I mean by "factually believe." That's true regardless of what day you take it to be: if you know (or just factually believe) it's Wednesday, believing it's Wednesday is not a choice; if you factually believe it's another day, you couldn't choose to switch to a factual belief that it's Wednesday, and if you have no idea what day it is, you couldn't induce in yourself a <u>factual belief</u> that it's Wednesday by direct

voluntary control (though you could either voluntarily form a <u>guess</u> or look for evidence that would make up your mind for you).

19. Van Leeuwen & van Elk (2019). Since this was a coauthored paper, we didn't make use of my idiosyncratic terminology of "religious credence," though van Elk is sympathetic to my views. Here, however, I adjust the terminology to stay consistent with the rest of this book.

20. People often form personal credences by interpreting low-level intuitive experiences *in light of* their general credences (where the general credences are mostly learned from one's surrounding culture).

21. For contemporary research on conversion that points in this direction, see Lofland & Stark (1965), Straus (1979), Greil (1977), Bromley & Shupe (1979), Long & Hadden (1983), Richardson (1978), Stark & Finke (2000), and Granqvist (2003).

22. To be more precise, the constraint is different: your factual beliefs about topics like what your address is are constrained by evidence (perceptual, etc.), but one's general religious credences are constrained not by evidence (as I argue below) but by social pressures from belonging to a religious group. Furthermore, insofar as there are social pressures to "believe," that's indirect evidence that general religious credences *are* voluntary in the relevant senses since it makes less sense to have normative social pressures for people to do things that they cannot voluntarily control. See Alston (1988) for compelling arguments in this direction.

23. Luhrmann (2012: 7); my italics.

24. In Dennett's (2006) terms, they "believe in belief."

25. Luhrmann (2012: 123); my italics.

26. This may not be true for basic actions like raising one's arm. But that qualifier is immaterial to the arguments in this subsection.

27. Luhrmann (2020b: 1); Luhrmann's italics.

28. Boyer (2013) makes a similar argument.

29. And though reminders for less frequently used factual beliefs are employed often enough, one does not need a "reminding" that pervasive things like electricity and money exist; one acts as if those are real seven days a week, in contrast to how most Christians act in relation to (their idea of) God. Note also that the idea of "strengthening" belief can be interpreted in different ways, as I point out in Chapter 8 and in Van Leeuwen (2022).

30. Bialecki (2017: 158).

31. Of course, according to official doctrines, God is supposed to be omnipresent. But if we set theological correctness aside and notice how Vineyard members (and other Christians) actually speak, we'll notice that they talk about God "really showing up" at a given service and represent God as being more present in church than elsewhere. Against that backdrop, it is highly significant that demons only seem to afflict people in situations in which God is supposed to be present because

such compartmentalization can't sensibly be explained by the *contents* of the relevant "beliefs."

32. Bialecki (2017: 155); my italics.

33. Gilkes (1980: 39).

34. This first expectation is often not explicitly articulated because "theologically correct" doctrine has it that God is present at all places and times (Barrett & Keil, 1996; Barrett, 1999). But that goes hand in hand with the fact that the compartmentalization of religious credences cannot be explained as a function of the *contents* of the religious beliefs, contrary to what Levy (2017) suggests.

35. Evans-Pritchard (1965: 88–89).

36. This is part of why I call all cognitive attitudes that are not factual beliefs "secondary": in representing practical setting, factual beliefs are conditions for the possibility of their existence, but not vice versa.

37. Again, in more technical idiom, we'd say *antisymmetric* to allow for reflexivity.

38. As I pointed out in the last chapter, people often misunderstand this third point: it doesn't imply that the process of imagining is never used in figuring out new factual beliefs, since imagining that *p* can be used to form a new factual belief that *p*, when it is realized that what one imagined (namely *p*) comports with other factual beliefs that one already has. In this case, the imagining that *p* is not *taken as given* for the sake of inferring new factual beliefs; rather, it is an exercise that helps one figure out what the consequences are *of one's factual beliefs*. So this is an example of factual beliefs actually governing themselves, not of imaginings governing factual beliefs.

39. Note that (1) by itself already implies lack of widespread cognitive governance since, if a certain cognitive attitude does *not* govern factual beliefs, its governance is not widespread.

40. And Bialecki here is inspired by Gilles Deleuze (1988).

41. Bialecki (2017: 169); my italics.

42. Bialecki (2017: 182).

43. Bialecki also points out that the most dramatic miracles (astonishing healings, etc.) tend to be reported second- and thirdhand (*I knew someone who said they knew a person who . . .*) and are often located in developing parts of the world where they would be hard to verify. This seems to represent a compromise between wanting to embrace the most impressive charismata and wanting not to embrace things that can be proved false.

44. Barrett (2001).

45. Cf. Rödlach (2006) on what AIDS victims in Zimbabwe are willing to pray for.

46. Rey (2007).

47. Luhrmann (2012: 95).

48. Bialecki (2017: 141).

49. Of course, people's acting on the religious credence layer (the transparency, in the metaphor) may cause them to *do* things that cause them to have different

factual beliefs from those they would have had (e.g., factual beliefs about which page in the hymnal a certain hymn is on). But that does not amount to being cognitive governance in the defined sense.

50. For more discussion of this issue, see Chapter 8, where I discuss Stigall's (2018) "incoherent triad" involving faith, belief, and evidence.

51. Luhrmann (2012: 299); my italics.

52. See Ichikawa (2020) for a persuasive analysis of faith that highlights how it involves epistemic risk. Note also that what I say here coheres with Kierkegaard's view in *Fear and Trembling*—more on which comes in Chapter 8.

53. Luhrmann (2012: 143).

54. Luhrmann (2012: 301).

55. Brahinsky (2020).

56. Sperber (1982: 171).

57. Luhrmann (2012: 316).

58. Nor can such cases be analyzed simply as a lower "degree" of belief. I deal with this in Chapter 8 along with other alternate views.

59. Bialecki (2017: 171–172).

60. Bialecki (2017: 171–172).

61. Bialecki's observation that it's only "young white men" who try to argue from evidence when they explain their departure is worth dwelling on. It suggests that background social norms and stereotypes that various demographic segments apply to themselves play a role in how they verbally justify leaving the church. This is speculative—and Bialecki does not proffer an explanation of the pattern in question—but I find it likely that young white men, more than other groups, identify with the norm of being able to "win" verbal arguments, which is why they are more likely to give a lawyerly gloss on their reasons for leaving the church, even though the deeper driving reasons are social and emotional.

62. One complexity here is that religious practitioners, including Vineyard members, often *talk* as if various events are "evidence" for their "beliefs." I refer to this as The Evidence Game, which is an *extension* of the identity-constituting make-believe that is characteristic of religious practice. See Van Leeuwen (2017b) for more on this. Generally, the fact that one might cite "evidence" as support for an attitude does not mean that that attitude is evidentially vulnerable in the sense that contrary evidence *extinguishes* it. One can cite evidence for the contents of one's imaginings, for example, but that does not mean imaginings are evidentially vulnerable in the relevant sense.

4. EVIDENCE AROUND THE WORLD

1. Huizinga (1938/1949).

2. Henrich, Heine, & Norenzayan (2010). See Henrich (2020) for a comprehensive development of the view.

3. To be clear, this is in no way suggested by Henrich, Heine, & Norenzayan (2010). Their topic is not religious "belief." Still, their notion of WEIRD is convenient for articulating a view to which many anthropologists might well subscribe.

4. Evans-Pritchard (1937/1976: 194).

5. Horton (1967: 155).

6. Toren (2007: 307–308).

7. Taylor (2007).

8. Legare & Gelman (2008: 617).

9. Legare et al. (2012).

10. Busch, Watson-Jones, & Legare (2017).

11. Legare & Shtulman (2018: 416).

12. Watson-Jones, Busch, & Legare (2015).

13. For more on this, see Qian's ancient history *The First Emperor* (trans. Dawson, 94 BCE/2007).

14. For academic corroboration of Achebe's portrayal of Igbo religion, see Ezenwa (2017) and Onyibor (2016).

15. Achebe (2013); quotations from pp. 19, 42, and 133, respectively; my italics.

16. Taves (2016).

17. Boyer (2001: 268–269); my italics.

18. Hines (2003).

19. If we look at church fathers, we see that the issue of forcing conversion—and, relatedly, the voluntariness of "belief"—was a matter of active dispute. To simplify matters greatly, it appears that both Augustine (Letter 93) and Aquinas (*Summa Theologiae*, Second Part of the Second Part, Question 10) thought that "belief" was to some extent up to the will, which seemed to provide them with some justification for advocating physically forceful incentives to bring unbelievers, pagans, Jews, and heretics (back) into the Christian faith. I suspect that their voluntarism, such as it was, was in part driven by the fact that the kind of "belief" they thought most extensively about was religious "belief." I doubt they would have endorsed the idea that one could decide to believe it was Wednesday when it was really Friday.

20. References: On King Clovis I, see Wood (1993); on Rollo the Viking, see Mark (2018); on Rolexana, see Lewis (2022); and on Tony Blair, see Bates (2007).

21. Luhrmann (2012: 70).

22. One other kind of incentive that may lead to conversion is money. According to anthropologist Hema Tharour (personal communication, 2016), some religious organizations have converted people to Christianity in India by promising them $50 and a hot meal on Sundays.

23. Astuti (2007: 241).

24. Astuti (2007: 234).

25. Nadeem Hussein raised this objection when I was first drafting Van Leeuwen (2014a) and discussed it with him.

26. Astuti (2007: 242); my italics.

27. Bratman (1992).

28. Astuti & Harris (2008: 734).

29. Harris & Giménez (2005); Watson-Jones et al. (2017). If we attempt to interpret these results with a One-Map Theory, we'll have to say that people who were studied constantly *changed their minds* about whether the deceased continue to have mental lives (now they do; now they don't; now they do . . .). And they would change their minds from context to context without evidence since the cues of a religious context do not typically encode any evidence about the metaphysics of the afterlife. That interpretation, on my view, lacks independent motivation. Why would what the Vezo think about the afterlife constantly oscillate in absence of new evidential inputs? A better interpretation is supplied by the Two-Map Theory. People have factual beliefs about what happens when people die, and they have religious credences: being in a sacred setting toggles on the religious credences, which represent the deceased as living-yet-invisible psychologically rich beings. The deceased, to use Bialecki's phrase, are thus "double coded."

30. Evans-Pritchard (1965: 8).

31. This caution, I think, should be heeded by defenders of One-Map Theories, such as Boudry & Coyne (2016a, 2016b), who, by focusing only on the verbal and nonverbal *religious* behaviors of religious people, conclude that those people think religiously all the time—hence, they miss the compartmentalization of religious credence.

32. Legare & Gelman (2008); quotations from pp. 636, 632, and 636, respectively; my italics.

33. Duhaime (2015). See Shariff et al. (2016) for an overview of this kind of research. Such studies on moral behavior may seem only to suggest compartmentalization of religious "beliefs" with *normative* contents, whereas I made it clear in Chapter 1 that my main concern here is with cognitive attitudes that are descriptive (they describe how the world or some portion of it *is,* rather than how it *should be* or how one *should* behave). But the studies are suggestive of the compartmentalization of descriptive religious cognitive attitudes as well. That's because much of the morality of religions with Big Gods, as Norenzayan (2013) puts it, is encoded in religious credences about what God approves of and, crucially, what God is inclined to punish. And contents like *God approves of X, God disapproves of X,* and *God punishes those who do X* are descriptive: they describe certain characteristics of the deity. The point is that credences with descriptive contents like *that* are compartmentalized.

34. Edelman (2009: 217–218). Importantly, it is not the *contents* of the "beliefs" about what God does or doesn't approve of that explains this instance of the Sunday Effect since there is nothing in the contents of Christian doctrine that suggests that pornography is acceptable on other days. A better explanation is that the compartmentalization of the *attitude* of religious credence explains the differential

pattern of behavior: since religious credences are more likely to be activated for guiding behavior on Sunday due to practical setting dependence, the idea that God disapproves of pornography is more likely to limit purchasing on that day but not others, as Edelman observes. This is a general problem for the content-based approach: it has difficulty explaining compartmentalization.

35. Malhotra (2010: 140).

36. In fairness to Fortes, he may well have been probing for a response and found this a useful provocation; his assumption, in other words, could well have been made with a certain intellectual distance.

37. Chaves (2010: 2–3); my italics in the first passage; Chaves's italics in the second.

38. If the "instrumental-looking" actions are not actually instrumental, you might wonder why people do them. I address that question in Chapter 6.

39. The qualification to this generalization is that people are more than willing to pray for impossible things when they are in mortal danger and no practical options are available. See Alexander Rödlach's (2006) study of the AIDS crisis in Zimbabwe for compelling examples of this tendency.

40. Atran (2002: 87).

41. By way of contrast, if you *factually believed* that one entity could transform into another, you would certainly worry that in eating the other, you might be eating the one.

42. Some people, of course, bite the bullet and grant the incest. But (i) this seems to be more the exception than the rule, and (ii) I think the reason for this is not that they factually believe the content but because the bullet-biters are engaged in a certain form of costly signaling that involves high epistemic costs, as I discuss in Chapter 6.

43. Barrett & Keil (1996: 224).

44. Barrett (1999).

45. Barrett (1999: 327).

46. Notably, we hear an echo of the famous Barrett and Keil study in a passage from Luhrmann's ethnography in which she discusses how Vineyard members talk about God "showing up" to a church service. "They say things like 'God really showed up today,' and then they distinguish between times when they felt that God was present and times when they did not feel his presence, although then they add quickly that he is always there" (6). The Barrett and Keil study and ensuing theory helps make sense of the psychology of such a paradoxical-seeming position: Vineyard members have an *intuitive conception* of God as a limited agent who isn't everywhere, and they have a *theologically correct* conception. The intuitive conception is the one that shapes their experience of the church service at which God either does or doesn't "show up." And the theologically correct conception—the one that people rely on to avoid heresy—is the one that leads them to add "quickly

that he is always there." If this is right, then the present point about what Barrett and Keil's study implies about cognitive governance carries over to the Vineyard: their theologically correct "beliefs" lack cognitive governance—both over what one imagines about God and even over many other religious credences.

47. He didn't use the word "evidence," but it was clear that the events he told me about were meant to have the effect of rational persuasion, as producing evidence typically does.

48. Festinger, Riecken, & Schachter (1956).

49. Because the years in many computer databases were represented with only two digits, the systems were expected to get confused between dates in the 1900s and 2000s, with ensuing pandemonium.

50. The most astonishing recent example of such persistence of "belief" even in the face of predicted dates coming and going with none of the predictions coming true occurs among adherents of the recent QAnon conspiracy theory. That conspiracy theory suffered a litany of embarrassments in terms of predicted dates passing without incident (e.g., March 4, 2021, was the predicted date for Donald Trump to resume office), with many followers still "believing" the theory nonetheless.

51. Formerly and at the time of the incident in question, it was Edmonds Community College.

52. Lyke (2003).

53. One might think that this just shows that Mormons do not accept DNA evidence in general, but they do. Ancestry.com, for example, was founded by Mormons so that they could trace ancestry to baptize ancestors who hadn't been baptized. So does DNA count as evidence of ancestry for Mormons or not? It seems that it does, except for when it does not.

54. The latter example is from personal communication with Felicity Aulino, an anthropologist who focuses on Thai Buddhism.

55. Appiah (2019: 37–38).

56. Clegg et al. (2019); Davoodi et al. (2018); Guerrero, Enesco, & Harris (2010).

57. Guerrero, Enesco, & Harris (2010: 146–147).

58. Liquin, Metz, & Lombrozo (2020).

59. Cf. Woolley & Cornelius (2017).

60. Sauvayre (2011).

5. TO "BELIEVE" IS NOT WHAT YOU "THINK"

1. Of course, we already have some reason to think that they do. Recall the line of Luhrmann's Vineyard informant: "I don't believe it, but I'm sticking with it. That's my definition of faith." This already implies awareness of different sorts of "belief." The question for this chapter, then, is how widespread we should take such awareness to be.

2. For a useful overview of the philosophical and psychological literature on theory of mind and social cognition, including the false belief task and much else, see Spaulding (2018).

3. Confirmed by Alison Gopnik (personal correspondence).

4. Hence the title of Levy's (2017) article: Religious Beliefs Are Factual Beliefs. Levy (2018: 821) also argues against my view by claiming that it entails that people are routinely "mistaken" about their own mental states. But my view would only have that implication if it were true that laypeople used "believe" and "belief" just like philosophers do, which this chapter shows to be false.

5. See Heiphetz, Landers, & Van Leeuwen (2021) for the full research report, including methods, participant demographics, statistical analyses, and further discussion of alternate hypotheses.

6. The standard measure in corpus linguistics of whether one word is associated with another in a significant way is called the *mutual information* (MI) score. The MI score is a comparison of how often two words occur near each other with how often they would be expected to occur near each other by chance. See Heiphetz, Landers, & Van Leeuwen (2021: n. 7) for the exact formula. Thanks to Ute Römer for help with developing the corpus study.

7. To be precise, we used the lemma *think* and the lemma *believe,* which capture all grammatical variants of the word in question.

8. See our supplemental materials for complete stimuli: https://supp.apa.org /psycarticles/supplemental/rel0000238/rel0000238_supp.html.

9. Of course, we only told participants that it was a sentence completion task where they were to find the word that sounds most natural. They didn't know anything of the sentence categories we had developed.

10. We randomized the sentence order to rule out order effects.

11. In fact, slightly the opposite pattern emerged, though not with an effect size that indicates anything interesting.

12. This is all the more striking when one considers that—as we learned in our corpus research—the word "think" occurs more than six times more often than "believe" in American English overall.

13. Note that by controlling for content, we also managed to rule out quite a number of tempting additional alternate explanations for our main effect; namely, those that appeal to features that depend on content: for example, whether a belief is verifiable, whether it is about things that are observable, and so on.

14. Another point, with us since Chapter 1, that gets illustrated by the fact that people can *believe* (in the religious credence sense) that Elvis is alive: once again, *anything can be sacralized.*

15. Of course, these are not the only two questions, but they are the most salient for purposes of this chapter. One other question that comes up often in discussions of this research is whether "know" gets used in relation to religious

items. See Heiphetz, Landers, & Van Leeuwen (2021: n. 13), which shows that participants used "know" in our Study 3 (free response) significantly less for religious items than for factual items. Van Leeuwen, Weisman, & Luhrmann (2021) also found less frequent use of "know" (or counterparts) for religious items in all their study locations except Ghana.

16. Luhrmann, ed. (2020a); Luhrmann et al. (2021).

17. Van Leeuwen et al. (2021).

18. The title (Does *Think* Mean the Same Thing as *Believe*?) of Heiphetz, Landers, & Van Leeuwen (2021) may—unfortunately, in retrospect—give the misleading impression that our main interest was semantics, but the body of the paper makes it clear that our main interest was people's ability to cognize religious credence and factual belief differently, which is an interest in the psychology much more than in semantics.

19. The obvious reason is that linguistic corpora aren't available in all the languages of interest.

20. Comprehensive stimuli in all languages, further analysis, and complete data sets are available here: https://osf.io/qy3js/.

21. For each study, stimuli were presented in one of two counterbalanced orders.

22. To be exact, the vignettes in this crosscultural study formed a proper subset of those in Heiphetz, Landers, & Van Leeuwen (2021), though lightly edited to streamline translation and ultimately facilitate comprehension. We basically continued with the vignettes that had the least cultural baggage, so the "aspirin" vignette pair was carried over from Heiphetz, Landers, & Van Leeuwen (2021) since that only assumes knowledge of aspirin, but the Elvis vignette pair was dropped.

23. Note that "Study 1" in Van Leeuwen et al. (2021) maps to "Study 2" in Heiphetz, Landers, & Van Leeuwen (2021), and so on since the crosscultural paper doesn't have a corpus linguistics component, which was "Study 1" in Heiphetz et al.

24. Dulin (2020); Dzkoto (2020).

25. See our Supplement (Study 3b) for discussion of a follow-up vignettes study we did in Ghana that did show a think/believe difference. This version of the vignettes study was conducted in English with university students, with the idea being that that would be one way of addressing the possibility that the nonresult of the earlier Study 3 in Ghana was due to deep cultural differences rather than mere comprehension issues. Since we did get an effect in this 3b—but only a small one—the issue is still very much open. Our Supplement can be accessed via the following URL: https://direct.mit.edu/opmi/article/doi/10.1162/opmi_a_00044/106928/To-Believe-Is-Not-to-Think-A-Cross-Cultural#supplementary-data.

26. Van Leeuwen et al. (2021: 98).

27. It's also worth noting that one finds something like the think/believe difference in quite a range of other languages as well: *denken* versus *glauben* (German), *penser* versus *croire* (French), *pensar* versus *creer* (Spanish), *düşünmek* versus *inanmak*

(Turkish), *chashav* versus *he'amin* (Hebrew), *luulla* versus *uskoa* (Finnish), *cabanga* versus *kholwa* (Zulu), and so on. What to make of these comparisons is up for debate, but it is clear that people across cultures and language groups are *capable* of tracking subtle distinctions in cognitive attitudes and try to find words to help them express the differences they track.

28. Dennett (2006) argues that there is also normative pressure against subjecting religions and religious beliefs to scientific investigation, which is the "spell" he's referring to in his title *Breaking the Spell*.

29. Straightforward corrections of this sort happen so quickly that they often seem not to even deserve the word "disagreement." But that just strengthens the point in question: straightforward corrections happen relatively easily with factual beliefs, but it's almost comical to think they might work with religious credences. (Consider: "You know people can't live in the belly of a whale for three days, right?" Someone who said that would either seem rude or clueless about the psychological dynamics of religious credence.)

30. See De Cruz (2018) for a work that treats disagreement on religious "belief" as more or less a special case of the kind of disagreement that epistemologists of disagreement study more generally, which treat matter-of-fact disagreements—those that don't involve people's identities—as paradigmatic. My point here is that that paints a misleading portrait of what religious credences are like as psychological states.

31. Heiphetz et al. (2013).

32. Examples were all of religious belief, and in their paper, they use "ideological belief" and "religious belief" as interchangeable.

33. See the portion of my next chapter that discusses dual direction of fit, which is one way of explicating the idea.

34. Liquin, Metz, & Lombrozo (2020). This is one of a line of recent papers to emerge from the Lombrozo lab that cohere well with the view put forward here and with my views about evidential vulnerability put forth in the previous chapters. See also Davoodi & Lombrozo (2022a, 2022b).

35. Williams (1973).

36. Just look at the use of "believe" in Goldman's (1970) classic *A Theory of Human Action*. It is one of the most frequent words in the book, and it generally designates what I call factual belief throughout.

37. Smith (1977/1998).

38. As pointed out earlier, Levy's (2018) argument against my view *depends* on assuming that laypeople use "belief" in the same manner as philosophers.

39. Gettier (1963).

40. Also, note that many of these debates fail to address Hume's Desideratum of distinguishing "belief" from other cognitive attitudes. Chalmers's (2011: 538) work on merely verbal disputes is also relevant here: "Likewise, instead of asking 'What is

a belief? What is it to believe?' and expecting a determinate answer, one can instead focus on the various roles one wants belief to play and say: here are some interesting states: B_1 can play these roles, B_2 can play these roles, B_3 can play these roles. Not much hangs on the residual verbal question of which is really belief."

41. Sperber (1996: 16).

42. The interesting twist that emerged in this chapter is that, if we look carefully, pretheoretic lay usage ("think" versus "believe") tracks the relevant phenomena (factual belief versus religious credence) *better* than unregimented philosophical "belief" use, which lumps together quite distinct cognitive attitudes.

6. IDENTITY AND GROUPISH BELIEF

1. Appiah (2019).

2. See D. Williams (2021) for a related game theoretic perspective on why absurd "beliefs" can be strategically useful.

3. Though my focus here is not on evolutionary origins, it is worth noting that the perspective I offer coheres with work on the cultural evolution of identity markers, such as ethnic ones. Employing the terms of McElreath, Boyd, & Richerson (2003: 123), I would put it as follows: Groupish beliefs are internal states that can generate *various* forms of "readily observable *marker trait,*" which facilitates cooperation with other group members. Insofar as Groupish beliefs play the role of generating symbolic markers, the pressure on them to have true and epistemically justifiable contents is basically none and may even be negative.

4. Note that the distinction here is similar to Konrad Talmont-Kaminski's (2013: 98) distinction between the "alethic" function of nonideological beliefs and the "non-alethic function" of ideological beliefs. Yet Talmont-Kaminski doesn't go as far as I think he should in terms of seeing the implications of his distinction for differences in cognitive attitudes. He writes: "At the same time, the non-alethic function of ideologies is parasitic upon the alethic function of other beliefs as ideologies must generally be believed to be literally true in order to motivate behavior." In my view, that sentence is inaccurate. All sorts of cognitive attitudes besides factual beliefs guide behavior, and it is quite certain that many ideologues throughout history have failed to factually believe the claims of their ideologies while acting them out (often in dramatic ways) nevertheless. What's needed, then, is an account of the cognitive attitude behind that behavior, which is where the notion of Groupish belief comes in.

5. Dretske (1983: 4).

6. Velleman (2000: 252).

7. See Spaulding (2018) for a recent overview.

8. It is true that many varieties of Judaism, for example, do not emphasize the having of "beliefs" in the way that most varieties of Christianity do, but that doesn't

mean that one can believe whatever one wants and still be Jewish in the religious sense: a belief in Vishnu might leave one's ethnic status as a Jew intact, but it would leave one's status as a member of most religious Jewish communities questionable. See Lanman (2008) for discussion.

9. The term "Groupish" is borrowed from Haidt (2012). The notion of Groupish Explanatory Role, as I develop it here, is my own.

10. Smith (1977/1998). In addition to laypeople, social psychologists also often associate the word "belief" with the Groupish Explanatory Role, which easily leads to cross talk when discussing "beliefs" with philosophers and other cognitive scientists, who may have Mundane beliefs in mind.

11. As, of course, Durkheim (1912/2008: Book III) argues.

12. Bulbulia (2004, 2012).

13. Appiah (2019: xiv).

14. One could have that *presentational* identity without even really loving Dylan's music.

15. Eliot (1915).

16. Cf. Appiah's (2018) *New York Times* piece "Go Ahead, Speak for Yourself."

17. Sartre's (1943/2003) waiter is an excellent example of someone acting out a social role identity. See Van Leeuwen (2011) for a way of thinking of such actions in terms of *semipretense.*

18. Thanks to Katherine Caldwell for calling the notion of imposed identity to my attention. The distinction here between imposed identity and group identity resembles Richard Jenkins's (2004) distinction between *categories* and *groups,* where the former are stipulated by the surrounding society and the latter embraced by the group itself.

19. Note that my approach to characterizing the psychology of group identity here is similar to the dimensional approach of Roccas et al. (2008), with the difference being that I spell out more dimensions and thereby offer a more granular conceptual space.

20. Dual direction of fit is clearly suggested in this passage from Liberman, Woodward, & Kinzler (2017: 556): "Social categorization differs from other forms of categorization in that *people tend to place themselves in a category* . . . leading them to be partial to members of their own group (ingroup) relative to those of other groups (outgroup) in terms of social preferences, empathic responding, and resource distribution" (my italics).

21. See Appiah (2019: 8–20) and Jenkins (2004: chap. 7) for further discussion of labeling. Note that an important qualification here is that the disposition in question may be audience-relative: I may be inclined to accept the label in front of some persons and not others. More generally, each of the features listed here can be further broken down in many complicated ways; my aim in this chapter is to hit the level of abstraction that allows us to say the most interesting things about the connections between group identity and "belief."

22. See Jenkins (2004: 82) for a related view on "criteria" for identity. On a different note, some criteria of inclusion can be off-loaded to "experts": I may only "know" a few criteria of inclusion, but the guru "knows" all of them.

23. For some identities, passing the litmus tests may not only be an *indicator* of criteria being satisfied; passing may be a criterion in and of itself. In principle, however, the two notions come apart. And it may be that litmus testing only tends to happen in harsher ideological conditions.

24. This notion of *habitus* originally stems from Bourdieu (1987), but it comes to me primarily via Appiah (2019: 20–25). An interesting line of support, however, comes from Watson-Jones, Whitehouse, & Legare (2016). They find that in-group ostracism among children leads to higher-fidelity imitation on the part of the ostracized. This may be seen as a conscious or unconscious way of clawing one's way back into the in-group through enhancing *habitus*.

25. Tajfel & Turner (1979). An interesting wrinkle here is that members of lower-status groups often can *in some sense* have a higher regard for members of higher-status out-groups. But that higher regard in terms of *prestige* is compatible with valuing members of one's own group more in other respects, such as worthiness of cooperation. See the quotation in my endnote 20 from Liberman, Woodward, & Kinzler.

26. A good illustration of such narratives can be found in Atran's (2010) *Talking to the Enemy,* where he discusses the "imagined kin" relations that members of religion and even terror cells posit in relation to one another. In most cases, there isn't actually a kinship relation between the respective people (and they know this); nevertheless, they Groupishly believe each other to be brothers and sisters. I think the fact that Atran uses "imagined" to describe such Groupish belief is deeply telling.

27. See Sperber (1982).

28. See Funkhouser (2017), Levy (2022: chap. 1), and Williams (2021) for alternate takes in the philosophical literature on the relation between beliefs and social signaling.

29. By "expresses," I mean that the symbolic action indicates that the person performing it *has* the Groupish belief in question; so, for example, writing about someone else's Groupish belief doesn't count as expressing it in the intended sense, so it is not symbolic action.

30. Sperber makes a comment in his (1975) *Rethinking Symbolism* that is relevant here. He notes in chapter 1 that one thing a theory of symbolism must explain is the ease with which anthropologists can identify it, even when they are new to a culture. Sperber's eventual explanation (p. 139) appeals to the "apparent gross irrationality" of the "manifestations" of "cultural symbolism." I have some sympathy with that explanation, but I don't think it is entirely right, since gross irrationality is not always apparent and many behavioral forms that do seem grossly irrational are not symbolic in the relevant sense, such as pretend play involving object substitution, and so on. I think my explanation is better: anthropologists are,

in fact, attuned to two things: (i) when behavior is representational and (ii) when it is expressive of group allegiance.

31. On the matter of group "beliefs" involving the positing of a sacred "essence," cf. Durkheim (1912/2008: 179): "The fundamental element of this religion is that members of the clan and the various beings represented by the totemic emblem are regarded as sharing the same essence."

32. One might respond that actions such as sacrifices also have practical goals attached to them, such as bringing about a good harvest. And if that is true, then there *is* pressure on the relevant Groupish beliefs to update in light of evidence. This is a tempting thought but recall from Chapter 4 that instrumental-looking actions are usually supplemental to practical actions in the direction of the relevant goal: one tends the fields in all the right ways to get a good harvest. Such accompanying practical actions leave the Groupish beliefs free to be false—and by extension, invulnerable to evidence—without a loss of success in practical affairs. And the regular accompaniment of parallel practical actions suggests that achieving practical goals is not typically the aim of symbolic actions anyway. Rather than being a mistaken means of getting a good harvest, the sacrifice is a successful means of signaling one's identity.

33. There are *many* complications here, but notice that several of the examples of inferential curtailment from Chapters 3 and 4 fit into this perspective. Calling an exorcist wouldn't actually fix the electrical problem with the coffee maker, praying without studying won't help someone ace their exam, concluding that the meat one is about to eat might be a person would limit one's food options: in each case, the inferential curtailment leaves forms of practical action available that would otherwise have been ruled out or neglected if religious credences had widespread governance.

34. Atran (2010).

35. Everyday parlance doesn't make the distinction so cleanly between fanaticism and extremism, but it will be useful to have these terms be clearer than they usually are. Note also that I don't make this distinction properly in Van Leeuwen (2014a and 2016a). In Van Leeuwen (2014a) in particular, I conflated what I call fanaticism here with extremism. The present account represents my more up-to-date thinking.

36. See Westover's (2018) account of her survivalist Mormon father's attitude toward hospitals.

37. This analysis, I think, also applies to Westover's (2018) account: at crucial times, when people were on the verge of death, they did go to the hospital.

38. Luhrmann (2012), for example, discusses how Vineyard members often talk with one another about their individual prayer experiences.

39. I address this apparent difference between my view and Durkheim's in Chapter 8.

40. Whether or not one succeeds in getting other group members to recognize one's membership *will* depend on matters beyond one's control. But the starting point is often choosing Groupish beliefs that set one up for success in the right ways.

41. Again, see Van Leeuwen and van Elk (2019) for more on the general/personal distinction.

42. Appiah (2019: chap. 2).

43. Norenzayan (2013).

44. Boyer & Baumard (2013, 2016).

45. See also Norenzayan & Shariff (2008).

46. Sosis & Alcorta (2003).

47. James (1902: 30). For the record, I think James offers a distorted picture when he says that "personal religion will prove itself more fundamental" and "should still seem the primordial thing." While some may have flashes of individual experience of "direct personal communication with the divine," as James puts it, these are generally rare and hard to come by (Luhrmann, 2012, 2020b). By way of contrast, the *communal* function of religion is far more pervasive and motivating, as the sociological literature on conversion, canvassed in Chapter 3, attests.

48. Dennett & LaScola (2010).

49. None of my published critics has put an objection exactly like this in print, though Boudry & Coyne (2016a) come close. Nevertheless, from a theory construction standpoint, it is one that is important to deal with. I thank Derek Baker for pushing me on this point.

50. Luhrmann (2012: 316).

51. Boyer & Liénard (2006: 816).

52. Taylor (2007: 5).

53. In fact, the word "sincere" would be odd if it were used in relation to most factual beliefs. If I asked whether you were sure you left your keys on the counter, it would be odd for you to respond, "I sincerely believe I left my keys on the counter." You would more likely just say you were sure or just that's where they are. "Sincere" connotes more of an emotional commitment (*to hold dear . . .*) than the dry epistemic commitment characteristic of factual belief. So if you see "sincere" before "belief," your first hypothesis should *not* be that straightforward factual belief is what's being referred to; Groupish beliefs are more likely referents of the phrase "sincere belief."

7. SACRED VALUES

1. *Euthyphro*, 6e-7a (in Plato, Fourth century BCE/1992); my translation.

2. And even if you did break down and take the wine, you would probably feel ashamed of yourself: rather than congratulating yourself for getting two bottles of

wine for the price of a couple scuffs on a dust jacket, you would regard your act as a sort of compromise of what's deeply important to you. In any case, your reaction is still of a different sort from your reaction in the dictionary case, which is what matters for present purposes.

3. By this, I do *not* mean to imply that there is a separate, distinct modular capacity for sacred values since it's clear that having sacred values implicates many psychological capacities that serve other purposes as well (like fear, imagination, and disgust). But it is clear that the attitude of sacred valuing activates those capacities in a coordinated enough way that we can call the collection of them operating in a certain way a system. So the main points I will assume in the background of this chapter are that (i) the sacred values system exists, and (ii) it is distinct from the system of ordinary utilitarian preference.

4. Alcorta & Sosis (2005).

5. Taves (2016), also discussed in Chapter 4.

6. Taves (2016: 42); my italics.

7. Walton (1990).

8. Kahneman & Tversky (1979).

9. Raz (1986); Tetlock et al. (2000).

10. Arguably, the existence of such an internal metric is a big part of what makes the use of money even possible.

11. Neil Levy (personal communication) points out that people can also show outrage in utilitarian exchanges, for example, if I were to offer you a ridiculously small amount of money for your car. But in that case, the outrage is not over the utilitarian exchange but over the fact that I have insulted *you*.

12. Cf. Tetlock et al. (2000: 855). One response to the claims in this paragraph goes like this: *surely, if there were enough money on the table (say, a billion dollars!), it would get people to move.* This answer is *not necessarily.* Scott Atran (personal communication) points out that one thing that sacred values can bring about is "blindness to exit strategies," which means that people will often continue pursuing what they value as sacred even at the cost of losing *everything* (friends, family, etc.). And in cases where people do take the money, which certainly exist, it is not the sacred values system that sanctions the choice; rather, the utilitarian values system *overrides* the sacred values system. The notion that the two systems can be in competition for control of the agent's actions is well encapsulated in the notion of *temptation.* When one gives in, one will then often rationalize the choice by trying to frame it in a way that is acceptable to the sacred values system, which is sometimes possible due to the voluntariness of religious credence.

13. The notion of *attaching* will be spelled out in section 5; for now, the rough idea should be clear enough to proceed.

14. Additivity can be derived from the independence preference axiom; see Peterson (2009: chap. 8) for discussion of that axiom. Note that there are some

complications here that are immaterial to the discussion in the main text. Additivity is actually a special case of a more general property, which is the following: $u(x, y)$ is greater than $v(x)$ and greater than $v(y)$ if both of the latter terms are positive, less than the value of the positive constituent if one is positive and the other is negative, and less than both if both are negative. In other words, even if $u(x, y)$ is not strictly the *sum* of $v(x)$ and $v(y)$, it is still moved by each of them *in the direction of* their independent value when they are put together. The reason we might wish to appeal to this more general property for characterizing the utilitarian value system comes from examples like this: though the utility of two houses is better than one, it is not just the sum of the values of each house if you only had that one (thanks to Kenny Easwaran for the example). Still, since the sacred values system violates this more general property in light of incentive outrage, this point of contrast with the utilitarian values system still holds up even under more nuanced consideration. I stay with additivity in the main text in any case for ease of exposition.

15. Ginges et al. (2007). Note that "incentive outrage" is my term and not theirs, but I think it usefully describes the phenomenon they've unearthed.

16. Ginges et al. (2007: 7358).

17. This Ginges et al. (2007) study illustrates another point I've been trying to make. Among both the Israelis and Palestinians, there were also subsets of individuals that did *not* exhibit incentive outrage: individuals in these subsets responded more positively to the deals with incentives—as additivity would lead one to predict. So the very same thing (in this case, land) can be sacralized or not, *even by people who are on the same side of an issue.* What it is for a thing to be sacralized on the picture developed here, then, is for the person's sacred values system to attach to that thing, where being so attached implies that the sacred values system treats it as constitutively incommensurable with objects of utilitarian value (i.e., objects to which one's utilitarian values system attaches) and reacts with outrage both at the possibility that the thing might be violated and at any incentives to violate it. Just as metal plates under a cloth can be sacralized or not, so, too, can patches of land in the Middle East be sacralized or not.

18. Ginges & Atran (2013).

19. See Nemeroff & Rozin (2018) for a recent review of contagion psychology research.

20. Tetlock (2003: 321).

21. Why and under what conditions does it go one way versus the other? That could be the subject of a book in and of itself. The present point, however, is more general: contagion thinking is active in sacred valuing far above the level at which it is active in utilitarian valuing.

22. On the role of punishment in maintaining group cooperation, see Henrich (2009), Sheik, Ginges, & Atran (2013), and Tetlock et al. (2000)—among many others.

23. Cf. Whitehouse (2018). This is another way that the sacred values system diverges from the utilitarian values system. People flock around preferences of any sort, but a sacred value plays a constitutive role in group identity that a preference for one kind of toaster over another does not. If utilitarian values were the values that formed part of one's group identity, group loyalty would *merely* depend on convenience and calculation of costs versus benefits, contrary to what we see.

24. Note that even *if* a determined decision theorist found *a* way of using rational choice theory to model actions motivated by sacred values, that would not undermine my distinct systems claim: it would merely show that the sacred values system *also* can be modeled with decision theoretic tools. In other words, conflicts between the intuitive theologian and the intuitive economist would still arise, even if the intuitive theologian can be modeled *somehow* in decision theoretic terms. I doubt that the attempt to describe the intuitive theologian in those terms would yield insight, but even if it did, it would still not undermine the main points of this chapter.

25. Some of the most interesting presentations of Kahneman's view come from Michael Lewis's (2016) *The Undoing Project*.

26. Evans-Pritchard (1965: 88–89), quoted also in Chapter 3.

27. Tetlock (2003: 322).

28. Accessed August 5, 2022, https://worldpopulationreview.com/country-rankings/abortion-rates-by-country.

29. Boyer (personal correspondence).

30. Saunders (2014).

31. Religious studies scholars debate exactly how to distinguish *natural* from *supernatural,* given that the lines between them *seem* to differ by culture. But all that's needed here is just *that* people in many different cultures track some such difference, even if what the difference amounts to isn't exactly the same everywhere.

32. This notion is inspired by John Perry's notion of detached mental files; for more, see Perry (2001b) and Van Leeuwen (2012).

33. Luhrmann (2012: chap. 8) points out that actual perceptual experiences of this kind (as opposed to mental imagery) are relatively rare.

34. Luhrmann (1980, 2001a, 2020b); Van Leeuwen & van Elk (2019).

35. As Liquin, Metz, & Lombrozo (2020) would lead us to expect: religious people tolerate mystery for their religious ideas in ways that people generally don't for factual beliefs.

36. It can't be overstated how different detached and linking religious credences are. Two people of the same religion can have the same detached religious credences—doctrines and stories—but if their linking religious credences are different, their religious behaviors in the physical world of here and now will be entirely different. Linking religious credences make possible *bodily* interactions—not just verbal expressions—with entities that have a role in the superordinary narra-

tives, and they make possible sensory experiences that can be internally categorized as being experiences *of* the supernatural. And since sensory experiences are primary when it comes to the generation of strong emotional states (Zajonc, 1984), linking religious credences enable mundane objects of perception to trigger religious awe, fear, wonder, love, and so on as if those mundane objects were the superordinary entities themselves.

37. Pierce (2017).

38. Astuti (2007).

39. See Goldman (1970) for an explanation of how actions of one type can conventionally generate actions of another, including representational actions.

40. One question I've received more than once by readers of drafts of this chapter is this: *How does the sacred values system you posit relate to whatever psychological processes produce moral values?* The question has various formulations that more or less get at the same issue. Here's another formulation: *How do sacred values of the religious sort relate to moral values in general?* I have an answer to this, but I first freely admit that I do not have a firm stance on the matter: I am quite sure that a sacred values system in something like the form I've described is worth positing, and it will be further work (philosophical and empirical) to determine how *it* relates to moral values generally. With that qualification out of the way, here is my view: terms like "moral values" and "moral judgment," when used in reference to psychological states that people actually get into, can refer to heterogeneous psychological kinds; for example, the psychological states of the effective altruist that count as her "moral values" are likely quite different from the psychological states of a conservative combat veteran that count as his "moral values," as a wealth of empirical research on such matters suggests. So the sacred values system I posit here should be regarded as *one type* of psychological kind that could be dubbed as "moral values" of a sort, though there certainly are others as well. There is much more to say here, but let this endnote serve as a sufficient peek inside Pandora's Box for now.

8. THE PUZZLE OF RELIGIOUS RATIONALITY

1. Dennett (2006).

2. Much more could be said on what epistemic rationality amounts to, but I rest with a minimal characterization here since it is clear that The Puzzle in question would arise for *any* fair characterization of epistemic rationality.

3. Sperber (1975).

4. Evans-Pritchard (1937/1976).

5. Kierkegaard (1843/2013).

6. Of course, if Kierkegaard's thesis is right, the book of Genesis already made the point through the story of Abraham and his "belief" in a God who would

have him sacrifice his son Isaac. New Testament texts point in the same direction. In the story of Doubting Thomas (John 20: 24–29), Jesus admonishes Thomas to "believe" without evidence, and Hebrews 11:1 seems to advise the same suspension of ordinary standards of evidence.

7. Stigall (2018).

8. Lombrozo (2014) suggests the attitude approach as a solution to a related problem in the following blog: https://www.npr.org/sections/13.7/2014/10/20/357519777 /are-factual-and-religious-belief-the-same.

9. Van Leeuwen (2013). Of course, imagination is not *entirely* unconstrained (Kind, 2016), but it is clear that it is far less constrained than factual belief.

10. I should add that there is far more that has been written than I can possibly mention here, by philosophers and anthropologists, on how it is possible to both be rational and maintain religious ideas. To indicate some further resources: Wilson (1974) is an excellent interdisciplinary collection; Winch (1964), which appears reprinted in Wilson's collection, has been particularly influential; and Tambiah (1990) explores different theoretical positions on the apparent opposition between magical/religious ideas and rationality that emerged in nineteenth- and twentieth-century anthropology. I hope that the taxonomy of solutions I offer here will be a useful tool for classifying different positions in that literature and that my Distinct Attitude Solution is a worthwhile contribution to it.

11. Dawkins (2006).

12. Freud (1927: 43).

13. McKay (2004: 10).

14. Note, however, that McKay has recently walked back his view from 2004. See McKay & Ross (2021).

15. For a wide-ranging study, see Zimmer et al. (2019).

16. This is not to say that people *never* have religious delusions—just that this is not the typical case. See McCauley & Graham (2020) for a nuanced discussion. I should also add that the discussion here would be complicated by an "acceptance" model of delusion (e.g., Dub, 2017), but the main point would still go through: the vast majority of religious people are *not* delusional in the way that, say, a schizophrenic person is. Thanks to Olivia Bailey for a discussion of this issue.

17. Levy (2019, 2022).

18. Harris (2012); cf. Mercier (2019: chap. 4).

19. There is currently a lively debate in the psychology of climate change denial on the role that knowledge versus ignorance of the relevant science plays in promoting or attenuating climate change denial. The current arguments don't hang on that debate. But interested readers should consider the following: Kahan et al. (2012), Kahan (2015), Ranney & Clark (2016), and Weisberg et al. (2021).

20. See also Mercier (2019) for a spirited and excellent defense of the view that people are *not* in general particularly gullible. Rather, we have evolved "open

vigilance mechanisms" that serve to help us differentiate trustworthy from untrustworthy information sources.

21. Festinger, Riecken, & Schachter (1956).

22. Again, see also Luhrmann (2012, 2018, 2020b) and Boyer (2013).

23. Most people associate Anselm with the metaphysical arguments of his *Proslogion,* which defends the existence of God as "that than which nothing greater can be conceived" (*id quo nihil maius cogitari potest*). His *Cur Deus Homo,* however, is more pertinent here since it is specifically a work of Christian apologetics.

24. Some very recent work that takes a cognitive evolutionary perspective arguably also falls under the Adjust the Rationality Approach, even if the authors wouldn't put it in those terms. Hong et al. (forthcoming), for example, argue that acceptance of rainmaking rituals can be explained by overestimation of efficacy due to "statistical artefacts," which would put the relevant "beliefs" at least in the ballpark of epistemic rationality. And Lightner & Hagen (2022) argue that "anthropomorphic and other supernatural explanations" result from "well-designed cognitive adaptations, which are designed for explaining the abstract and causal structure of complex, unobservable, and uncertain phenomena" (from their abstract). Though I am unconvinced, these approaches are quite interesting and, if successful, may alleviate The Puzzle for *some* cases of apparently irrational religious belief. But they won't work for *all*: belief in florid details such as the animal heads of deities in various religious traditions, for example, won't be explained by such approaches, and religions supply *many* examples of such florid detail.

25. Gould (1997).

26. Durkheim (1912/2008: 70); my italics.

27. Durkheim (1912/2008: 170–171).

28. I find Sperber's (1982) terminology in his "Apparently Irrational Beliefs" most perspicuous, so I focus on that, but where useful, I also draw in elements of other papers. (I find that Sperber's [1997] terminology lends itself to being confused with the standard System 1/System 2 distinction from Kahneman (2011) and others, with which it is *not* equivalent. So the [1982] terminology is more useful.)

29. The phrase "validating context" is from Sperber (1997).

30. Both phrases are key terms in Sperber (1982).

31. Sperber (1982).

32. Personal correspondence.

33. Relatedly, Langland-Hassan (2020) puts forth a parsimony argument for doing without a distinct cognitive attitude of imagining. Such arguments strike me as penny-wise (achieving little by way of parsimony) and pound-foolish (necessitating costly increases in the complexity of contents one must posit); see Van Leeuwen (2011).

34. Boyer (2001: 66).

35. Wittgenstein (1931/1993).

36. See Cottingham (2009) for dissent.

37. Wittgenstein (1931/1993: 119); italics in original.

38. Wittengstein (1931/1993: 68); italics in original.

39. Compare this point to "the religious congruence fallacy" discussed by Chaves (2010) and in Chapter 4 of this book.

40. Glock (1996: 321). See Glock's *Dictionary* for a key to the abbreviated references.

41. This anecdote is from Peter Asher, who lived with McCartney for a time, via his now-popular Sirius XM radio show, "From Me to You," on The Beatles Station on Sirus XM.

42. Luhrmann (2018, 2020b); Boyer (2013).

43. Sosis & Alcorta (2003).

44. Though we granted that *sometimes* contents are semipropositional.

45. Dennett (2006). Dennett's view of belief comes out in several influential publications (1971, 1978, 1987). What it is to have an intentional state, like belief, on Dennett's view is a matter of being predictable and explainable using the intentional stance, where taking the intentional stance involves *assuming* the general rationality and coherence of an intentional agent. It is thus fair to say that Dennett uses the term "belief" for a state that is characteristically rational, similarly to how I use "factual belief."

46. Dennett (2006: 227).

47. Lanman's (2020) blog is here: https://tif.ssrc.org/2020/01/31/belief-lanman/.

48. See also Lanman (2008).

49. Boudry & Coyne (2016a: 602).

50. Jong (2018).

51. I develop this distinction in Van Leeuwen (2022).

52. Atran (2010).

53. Alston (1996) and Audi (2008) both posit a distinct cognitive attitude of *acceptance* as an attitude type people can take toward ideas of their respective religions. Neither of them, however, directly address The Puzzle in the way that I do here (though they clearly have the machinery to do so). See also Rey (2007) for the interesting view that many religious people (at least in Western societies) are self-deceived with respect to their religious myths and doctrines.

EPILOGUE: THE PLAYGROUND EXPANDED

1. This paragraph was inspired by a conversation I had with Barry Lam in New York in the summer of 2017, right after we recorded my audio for the "Creed and Credences" episode of Hi-Phi Nation. It aims at what I consider to be the mainstream in analytic epistemology. Given that target, I take my points to be fair. That said, there is a subliterature in philosophy of religion that addresses the sort of ques-

tion I put on the table in this paragraph. In particular, there is a smallish camp of theorists who advocate what are called "nondoxastic accounts of faith," which bear a structural resemblance to my account, even though they contain far less psychological detail. And some of those thinkers in that nondoxastic camp (e.g., Audi, 2008) do indeed pose the question of what it would take to rationally support the "acceptance" component of religious faith. See Buchak (2017) for a helpful survey of that literature, which I regard as a useful start in raising the sorts of questions I think epistemology can now raise much more sharply in light of the theory I advocate here. I myself (Van Leeuwen, 2014a) proposed two normative principles, Balance and Immunity, for cognitive attitudes in general, which would thus include religious belief, and Balance is also suggested by Audi (2008). But I have come to regard the issue as more complicated than my earlier discussion suggests, so we should treat it only as a starting point in the relevant normative inquiry.

2. See, for example, Van Leeuwen (2007, 2008, 2009b). On the issue of cross talk involving the term "self-deception," see my (Van Leeuwen, 2010) *Philosophy Talk* blog on self-deception.

3. Recall also that this was one of the main points of the extended quotation from Luhrmann that I referenced in section 2.2 of Chapter 3 (Luhrmann, 2020b: 1). See also the ever-relevant "Belief in Belief" chapter from Dennett (2006: chap. 8).

4. The sort of self-deception posited here has some similarities to, and some differences from, the one that Rey (2007) posits in his well-known piece "Meta-Atheism." The key similarity lies in positing that the self-deceiving concerns what one's own "belief" states are like.

5. S. Harris (2004: x–xi); Harris's italics.

6. Darwall (1988: 424–425).

References

Achebe, C. 2013. *Arrow of God.* New York: Penguin.

Alcorta, C., & R. Sosis. 2005. Ritual, Emotion, and Sacred Symbols: The Evolution of Religion as an Adaptive Complex. *Human Nature* 16, no. 4: 323–359.

Alston, W. P. 1988. The Deontological Conception of Epistemic Justification. *Philosophical Perspectives* 2:257–299.

———. 1996. Belief, Acceptance, and Religious Faith. In *Faith, Freedom, and Rationality,* edited by J. Jordan and D. Howard-Snyder, 10–27. Lanham, MD: Rowman & Littlefield.

Alumona, V. 2003. Culture and Societal Institutions in Chinua Achebe's Things Fall Apart: A Critical Reading. *Journal of Humanities* 17:62–81.

Anselm. Eleventh–twelfth centuries/1998. *Anselm of Canterbury: The Major Works,* edited by B. Davies & G. R. Evans. Oxford: Oxford World's Classics.

Appiah, K. A. 2018. Go Ahead, Speak for Yourself. *New York Times,* Opinion, August 10.

———. 2019. *The Lies That Bind: Rethinking Identity.* New York: Liveright.

Aquinas, T. Thirteenth century/2006. *Summa Theologiae,* edited by T. Gilby. Cambridge, UK: Cambridge University Press.

Armstrong, D. M. 1973. *Belief, Truth and Knowledge.* Cambridge: Cambridge University Press.

Aronson, E., T. Wilson, & R. Akert. 2004. *Social Psychology.* 5th ed. Upper Saddle River, NJ: Pearson Education.

Astuti, R. 2007. What Happens after Death? In *Questions of Anthropology,* edited by R. Astuti, J. Parry, & C. Stafford, 227–247. Oxford: Berg.

Astuti, R., & P. L. Harris. 2008. Understanding Mortality and the Life of the Ancestors in Rural Madagascar. *Cognitive Science* 32, no. 4: 713–740.

Atran, S. 2002. *In Gods We Trust: The Evolutionary Landscape of Religion.* Oxford: Oxford University Press.

———. 2006. Sacred Values, Terrorism, and the Limits of Rational Choice. Paper presented to the US Interagency Meeting, Department of Homeland Security, April 2006.

———. 2010. *Talking to the Enemy: Faith, Brotherhood, and the (Un)Making of Terrorists.* New York: HarperCollins.

Atran, S., & R. Axelrod. 2008. Reframing Sacred Values. *Negotiation Journal* 24, no. 3: 221–246.

Atran, S., & J. Ginges. 2012. Religious and Sacred Imperatives in Human Conflict. *Science* 336, no. 6083: 855–857.

Atran, S., & A. Norenzayan. 2004. Religion's Evolutionary Landscape: Counterintuition, Commitment, Compassion, and Communion. *Behavioral and Brain Sciences* 27:713–770.

Audi, R. 2008. Belief, Faith, and Acceptance. *International Journal for Philosophy of Religion* 63, nos. 1–3: 87–102.

Augustine. 408/1887. Letter 93, "To Vincent." J. G. Cunningham, translated and revised by K. Knight. *Nicene and Post-Nicene Fathers, First Series, Vol. 1,* edited by Philip Schaff. Buffalo, NY: Christian Literature.

Aulino, F. 2020. From Karma to Sin: A Kaleidoscopic Theory of Mind and Christian Experience in Northern Thailand. *Journal of the Royal Anthropological Institute* 26, no. S1: 28–44.

Baillargeon, R. 2002. The Acquisition of Physical Knowledge in Infancy: A Summary in Eight Lessons. In *Blackwell Handbook of Childhood Cognitive Development,* edited by U. Goswami, 47–83. Hoboken, NJ: Wiley-Blackwell.

Barrett, J. 2001. How Ordinary Cognition Informs Petitionary Prayer. *Journal of Cognition and Culture* 1, no. 3: 259–269.

Barrett, J. L. 1999. Theological Correctness: Cognitive Constraint and the Study of Religion. *Method & Theory in the Study of Religion* 11, no. 4: 325–339.

Barrett, J. L., & F. C. Keil. 1996. Conceptualizing a Nonnatural Entity: Anthropomorphism in God Concepts. *Cognitive Psychology* 31, no. 3: 219–247.

Barrett, J. L., & J. A. Lanman. 2008. The Science of Religious Beliefs. *Religion* 38, no. 2: 109–124.

Bates, S. 2007. After 30 Years as a Closet Catholic, Blair Finally Puts Faith before Politics. *Guardian,* Politics Section, June 21.

Batson, C. D., J. H. Eklund, V. L. Chermok, J. L. Hoyt, & B. G. Ortiz. 2007. An Additional Antecedent of Empathic Concern: Valuing the Welfare of the Person in Need. *Journal of Personality and Social Psychology* 93, no. 1: 65–74.

Baumard, N., & P. Boyer. 2013. Explaining Moral Religions. *Trends in Cognitive Sciences* 17, no. 6: 272–280.

Bialecki, J. 2017. *A Diagram for Fire: Miracles and Variation in an American Charismatic Movement.* Oakland: University of California Press.

Bloom, P. 2015. Scientific Faith Is Different from Religious Faith. *Atlantic,* Science Section, November 24.

Bortolotti, L. 2005. Delusions and the Background of Rationality. *Mind & Language* 20, no. 2: 189–208.

———. 2015. The Epistemic Innocence of Motivated Delusions. *Consciousness and Cognition* 33:490–499.

Boudry, M., & J. Coyne. 2016a. Disbelief in Belief: On the Cognitive Status of Supernatural Beliefs. *Philosophical Psychology* 29, no. 4: 601–615.

———. 2016b. Fakers, Fanatics, and False Dilemmas: Reply to Van Leeuwen. *Philosophical Psychology* 29, no. 4: 622–627.

Bourdieu, P. 1987. *Distinction: A Social Critique of Judgment and Taste,* translated by R. Nice. Cambridge, MA: Harvard University Press.

Boyd, R. 1991. Realism, Anti-Foundationalism and the Enthusiasm for Natural Kinds. *Philosophical Studies* 61:127–148.

Boyer, P. 2001. *Religion Explained: The Evolutionary Origins of Religious Thought.* New York: Basic.

———. 2013. Why "Belief" Is Hard Work: Implications of Tanya Luhrmann's *When God Talks Back. HAU: Journal of Ethnographic Theory* 3, no. 3: 349–357.

Boyer, P., & N. Baumard. 2016. Projecting WEIRD Features on Ancient Religions. *Behavioral and Brain Sciences* 39:E6.

Boyer, P., & P. Liénard, P. 2006. Why Ritualized Behavior? Precaution Systems and Action Parsing in Developmental, Pathological and Cultural Rituals. *Behavioral and Brain Sciences* 29, no. 6: 595–613.

Brahinsky, J. 2020. Crossing the Buffer: Ontological Anxiety among US Evangelicals and an Anthropological Theory of Mind. *Journal of the Royal Anthropological Institute* 26, no. S1: 45–60.

Bratman, M. E. 1992. Practical Reasoning and Acceptance in a Context. *Mind* 101, no. 401: 1–16.

Bromley, D. G., & A. D. Shupe. 1979. "Just a Few Years Seem Like a Lifetime": A Role Theory Approach to Participation in Religious Movements. *Research in Social Movements, Conflicts, and Change* 2:159–185.

Buchak, L. 2014. Belief, Credence, and Norms. *Philosophical Studies* 69, no. 3: 285–311.

———. 2017. Reason and Faith. In *The Oxford Handbook of the Epistemology of Theology,* edited by W. J. Abraham & F. D. Aquino, 46–63. Oxford: Oxford University Press.

Bulbulia, J. 2004. The Cognitive and Evolutionary Psychology of Religion. *Biology and Philosophy* 19, no. 5: 655–686.

———. 2012. Spreading Order: Religion, Cooperative Niche Construction, and Risky Coordination Problems. *Biology and Philosophy* 27, no. 1: 1–27.

Busch, J. T., R. E. Watson-Jones, & C. H. Legare. 2017. The Coexistence of Natural and Supernatural Explanations within and across Domains and Development. *British Journal of Developmental Psychology* 35, no. 1: 4–20.

Chalmers, D. 2011. Verbal Disputes. *Philosophical Review* 120, no. 4: 515–566.

Chaves, M. 2010. Rain Dances in the Dry Season: Overcoming the Religious Congruence Fallacy. *Journal for the Scientific Study of Religion* 49, no. 1: 1–14.

Chomsky, N. 1957. *Syntactic Structures.* New York: Mouton de Gruyter.

Clegg, J. M., Y. K. Cui, P. L. Harris, & K. H. Corriveau. 2019. God, Germs, and Evolution: Belief in Unobservable Religious and Scientific Entities in the US and China. *Integrative Psychological and Behavioral Science* 53, no. 1: 93–106.

Clifford, W. K. 1877/1999. The Ethics of Belief. In *The Ethics of Belief and Other Essays,* edited by T. Madigan, 70–96. Amherst, MA: Prometheus.

Coleman, T. J. (host), & I. Visuri (guest). February 5, 2018. Autism, Religion, and Imagination [audio podcast episode]. The Religious Studies Project.

Cottingham, J. 2009. The Lessons of Life: Wittgenstein, Religion and Analytic Philosophy. In *Wittgenstein and Analytic Philosophy: Essays for P.M.S. Hacker,* edited by H.-J. Glock & J. Hyman, 203–227. Oxford: Oxford University Press.

Currie, G., & I. Ravenscroft. 2002. *Recreative Minds: Imagination in Philosophy and Psychology.* Oxford: Oxford University Press.

Darwall, S. L. 1988. Self-Deception, Autonomy, and Moral Constitution. In *Perspectives on Self-Deception,* edited by A. O. Rorty & B. P. McLaughlin, 407–430. Berkeley: University of California Press.

Davidson, D. 1963. Actions, Reasons, and Causes. *Journal of Philosophy* 60, no. 23: 685–700.

———. 1984. *Inquiries into Truth and Interpretation.* Oxford: Oxford University Press.

Davoodi, T., M. Jamshidi-Sianaki, F. Abedi, A. Payir, Y. K. Cui, P. L. Harris, & K. H. Corriveau. 2018. Beliefs about Religious and Scientific Entities among Parents and Children in Iran. *Social Psychological and Personality Science* 10, no. 7: 847–855.

Davoodi, T., & T. Lombrozo. 2022a. Varieties of Ignorance: Mystery and the Unknown in Science and Religion. *Cognitive Science* 46, no. 4: e13129.

———. 2022b. Explaining the Existential: Scientific and Religious Explanations Play Different Functional Roles. *Journal of Experimental Psychology: General* 151, no. 5: 1199–1218.

Dawkins, R. 2006. *The God Delusion.* New York: Houghton Mifflin.

D'Cruz, J. 2015. Rationalization as Performative Pretense. *Philosophical Psychology* 28, no. 7: 980–1000.

De Cruz, H. 2018. *Religious Disagreement.* Cambridge, UK: Cambridge University Press.

Deleuze, G. 1988. *Foucault,* translated by S. Hand. Minneapolis: University of Minnesota Press.

Dennett, D. C. 1971. Intentional Systems. *Journal of Philosophy* 68:87–106.

———. 1978. *Brainstorms: Philosophical Essays on Mind and Psychology.* Cambridge, MA: MIT Press.

———. 1987. *The Intentional Stance.* Cambridge, MA: MIT Press.

———. 2006. *Breaking the Spell: Religion as a Natural Phenomenon.* New York: Penguin.

Dennett, D. C., & L. LaScola. 2010. Preachers Who Are Not Believers. *Evolutionary Psychology* 9:122–150.

de Waal, F. B. M. 2016. *Are We Smart Enough to Know How Smart Animals Are?* New York: W. W. Norton.

Dretske, F. I. 1983. The Epistemology of Belief. *Synthese* 55, no. 1: 3–19.

Dub, R. 2017. Delusions, Acceptances, and Cognitive Feelings. *Philosophy and Phenomenological Research* 94, no. 1: 27–60.

Duhaime, E. 2015. Is the Call to Prayer a Call to Cooperate? A Field Experiment on the Impact of Religious Salience on Prosocial Behavior. *Judgment and Decision Making* 10, no. 6: 593–596.

Dulin, J. 2020. Vulnerable Minds, Bodily Thoughts, and Sensory Spirits: Local Theory of Mind and Spiritual Experience in Ghana. *Journal of the Royal Anthropological Institute* 26, no. S1: 61–76.

Durkheim, É. 1912/2008. *The Elementary Forms of Religious Life,* translated by C. Cosman. Oxford: Oxford World's Classics.

Dzkoto, V. 2020. Adwenhoasem: An Akan Theory of Mind. *Journal of the Royal Anthropological Institute* 26, no. S1: 77–94.

Edelman, B. 2009. Markets: Red Light States: Who Buys Online Adult Entertainment? *Journal of Economic Perspectives* 23, no. 1: 209–220.

Egan, F. 2014. How to Think about Mental Content. *Philosophical Studies* 170, no. 1: 115–135.

Eliot, T. S. 1915. The Love Song of J. Alfred Prufrock. *Poetry: A Magazine of Verse,* June 1915, 130–135.

Evans-Pritchard, E. E. 1937/1976. *Witchcraft, Oracles and Magic among the Azande.* Oxford: Oxford University Press.

———. 1965. *Theories of Primitive Religion.* Oxford: Oxford University Press.

Ezenwa, P. C. 2017. *The Value of Human Dignity: A Socio-Cultural Approach to Analyzing the Crisis of Values among Igbo People of Nigeria.* Doctoral diss., University of Würzburg.

Festinger, L., H. Riecken, & S. Schachter. 1956. *When Prophecy Fails: A Social and Psychological Study of a Modern Group That Predicted the Destruction of the World*. New York: Harper-Torchbooks.

Fodor, J. A. 1985. Fodor's Guide to Mental Representation: The Intelligent Auntie's Vade-Mecum. *Mind* 94, no. 373: 76–100.

Frankish, K. 2004. *Mind and Supermind*. Cambridge: Cambridge University Press.

Freud, S. 1927. The Future of an Illusion. In *The Standard Edition of the Complete Psychological Works of Sigmund Freud*, Vol. 21. London: Hogarth Press.

Friedman, J. 2013. Suspended Judgment. *Philosophical Studies* 162, no. 2: 165–181.

Funkhouser, E. 2017. Beliefs as Signals: A New Function for Beliefs. *Philosophical Psychology* 30, no. 6: 809–831.

Gelman, S. A. 2003. *The Essential Child: Origins of Essentialism in Everyday Thought*. Oxford: Oxford University Press.

Gelman, S. A., & J. D. Coley. 1990. The Importance of Knowing a Dodo Is a Bird: Categories and Inferences in 2-Year-Old Children. *Developmental Psychology* 26, no. 5: 796–804.

Gendler, T. S. 2003. On the Relation between Pretense and Belief. In *Imagination, Philosophy, and the Arts,* edited by M. Kieran & D. Lopes, 132–149. London: Routledge.

———. 2006. Imaginative Contagion. *Metaphilosophy* 37, no. 2: 183–203.

———. 2007. Self-Deception as Pretense. *Philosophical Perspectives* 21, no. 1: 231–258.

Gervais, W. M., & A. Norenzayan. 2012. Analytic Thinking Promotes Religious Disbelief. *Science* 336, no. 6080: 493–496.

Gettier, E. L. 1963. Is Justified True Belief Knowledge? *Analysis* 23, no. 6: 121–123.

Gilkes, C. T. 1980. The Black Church as a Therapeutic Community: Suggested Areas for Research into the Black Religious Experience. *Journal of the Interdenominational Theological Center* 8, no. 1: 29–44.

Ginet, C. 2001. Deciding to Believe. In *Knowledge, Truth, and Duty: Essays on Justification, Responsibility, and Virtue,* edited by M. Steup, 63–76. Oxford: Oxford University Press.

Ginges, J., & S. Atran. 2013. Sacred Values and Cultural Conflict. In *Advances in Culture and Psychology,* edited by M. J. Gelfand, C.-Y. Chiu, & Y.-Y. Hong, 273–301. Oxford: Oxford University Press.

Ginges, J., S. Atran, D. Medin, & K. Shikaki, 2007. Sacred Bounds on Rational Resolution of Violent Political Conflict. *Proceedings of the National Academy of Sciences* 104, no. 18: 7357–7360.

Ginges, J., S. Atran, S. Sachdeva, & D. Medin. 2011. Psychology Out of the Laboratory: The Challenge of Violent Extremism. *American Psychologist* 66, no. 6: 507.

Glock, H.-J. 1996. *A Wittgenstein Dictionary*. Hoboken, NJ: Wiley-Blackwell.

Goffman, E. 1959/2002. *The Presentation of Self in Everyday Life.* New York: Anchor.

Goldman, A. I. 1970. *A Theory of Human Action.* Englewood Cliffs, NJ: Princeton University Press.

Golomb, C., & R. Kuersten. 1996. On the Transition from Pretence Play to Reality: What Are the Rules of the Game? *British Journal of Developmental Psychology* 14, no. 2: 203–217.

Goodstein, J., & D. S. Weisberg. 2009. What Belongs in a Fictional World? *Journal of Cognition and Culture* 9, nos. 1–2: 69–78.

Gopnik, A. 2003. The Theory Theory as an Alternative to the Innateness Hypothesis. In *Chomsky and His Critics,* edited by L. M. Antony and N. Horstein, 238–254. Oxford, UK: Blackwell.

Gould, S. J. 1997. Nonoverlapping Magisteria. *Natural History* 106:16–22.

Granqvist, P., & B. Hagekull. 2003. Longitudinal Predictions of Religious Change in Adolescence: Contributions from the Interaction of Attachment and Relationship Status. *Journal of Social and Personal Relationships* 20, no. 6: 793–817.

Greil, A. L. 1977. The Modernization of Consciousness and the Appeal of Fascism. *Comparative Political Studies* 10, no. 2: 213–238.

Guerrero, S., I. Enesco, & P. L. Harris. 2010. Oxygen and the Soul: Children's Conception of Invisible Entities. *Journal of Cognition and Culture* 10, nos. 1–2: 123–151.

Guthrie, S. 1993. *Faces in the Clouds: A New Theory of Religion.* Oxford: Oxford University Press.

Haidt, J. 2012. *The Righteous Mind: Why Good People Are Divided by Politics and Religion.* New York: Vintage.

Harris, P. L. 2000. *The Work of the Imagination.* Oxford, UK: Blackwell.

———. 2012. *Trusting What You're Told: How Children Learn from Others.* Cambridge, MA: Harvard University Press.

Harris, P. L., & M. Giménez. 2005. Children's Acceptance of Conflicting Testimony: The Case of Death. *Journal of Cognition and Culture* 5, nos. 1–2: 143–164.

Harris S. 2004. *Letter to a Christian Nation.* New York: Vintage.

Harris, S., J. T. Kaplan, A. Curiel, S. Y. Bookheimer, M. Iacoboni, & M. S. Cohen. 2009. The Neural Correlates of Religious and Nonreligious Belief. *PLoS One* 4, no. 10: e7272.

Heiphetz, L., C. L. Landers, & N. Van Leeuwen. 2021. Does *Think* Mean the Same Thing as *Believe*? Linguistic Insights into Religious Cognition. *Psychology of Religion and Spirituality* 13, no. 3: 287–297.

Heiphetz, L., E. S. Spelke, P. L. Harris, & M. R. Banaji. 2013. The Development of Reasoning about Beliefs: Fact, Preference, and Ideology. *Journal of Experimental Social Psychology* 49, no. 3: 559–565.

Henrich, J. 2009. The Evolution of Costly Displays, Cooperation and Religion. *Evolution and Human Behavior* 30, 244–260.

———. 2020. *The WEIRDest People in the World: How the West Became Psychologically Peculiar and Particularly Prosperous.* New York: Farrar, Straus and Giroux.

Henrich, J., S. J. Heine, & A. Norenzayan. 2010. The Weirdest People in the World? *Behavioral and Brain Sciences* 33, nos. 2–3: 61–83.

Hines, J. 2003. The Conversion of the Old Saxons. In *The Continental Saxons from the Migration Period to the Tenth Century,* edited by D. Green & F. Siegmund, 299–314. Woodbridge, UK: Boydell.

Hong, Z., E. Slingerland, & J. Henrich. Forthcoming. Magic and Empiricism in Early Chinese Rainmaking—A Cultural Evolutionary Analysis. *Current Anthropology.*

Hood, B. M. 1995. Gravity Rules for 2- to 4-Year Olds? *Cognitive Development* 10, 577–598.

Horton, R. 1967. African Traditional Thought and Western Science. *Africa: Journal of the International African Institute* 37, no. 2: 155–187.

Huizinga, J. 1938/1949. *Homo Ludens: A Study of the Play-Element in Culture.* London: Routledge & Kegan Paul.

Hume, D. 1748/2000. *An Enquiry Concerning Human Understanding.* Oxford: Oxford University Press.

Ichikawa, J. 2020. Faith and Epistemology. *Episteme* 17, no. 1: 121–140.

Jackson, E. 2020. The Relationship Between Belief and Credence. *Philosophy Compass* 15, no. 6: 1–13.

James, W. 1902. *The Varieties of Religious Experience: A Study in Human Nature.* New York: Longmans, Green.

Jenkins, R. 2004. *Social Identity.* 2nd ed. London: Routledge.

Johnston, M. 1988. Self-Deception and the Nature of Mind. In *Perspectives on Self-Deception,* edited by B. McLoughlin & A. O. Rorty, 63–91. Berkeley: University of California Press.

Jong, J. 2018. Beliefs Are Object-Attribute Associations of Varying Strength. *Contemporary Pragmatism* 15, no. 3: 284–301.

Kahan, D. M. 2015. Climate-Science Communication and the *Measurement Problem. Political Psychology* 36:1–43.

Kahan, D. M., E. Peters, M. Wittlin, P. Slovic, L. L. Ouellette, D. Braman, & G. Mandel. 2012. The Polarizing Impact of Scientific Literacy and Numeracy on Perceived Climate Change Risks. *Nature Climate Change* 2:732–735.

Kahneman, D. 2011. *Thinking, Fast and Slow.* New York: Farrar, Straus and Giroux.

Kahneman, D., & A. Tversky. 1979. Prospect Theory: An Analysis of Decision under Risk. *Econometrica* 47, no. 2: 263–292.

Kierkegaard, S. 1843/2013. *Fear and Trembling and the Sickness unto Death,* translated by W. Louwrie. Princeton, NJ: Princeton University Press.

Kim, S., C. W. Kalish, & P. L. Harris. 2012. Speaker Reliability Guides Children's Inductive Inferences about Novel Properties. *Cognitive Development* 27, no. 2: 114–125.

Kind, A. 2016. Imagining under Constraints. In *Knowledge through Imagination,* edited by A. Kind & P. Kung, 145–159. Oxford: Oxford University Press.

Kuhn, T. S. 1962. *The Structure of Scientific Revolutions.* Chicago: University of Chicago Press.

Kyburg, H. E. 1983. Rational Belief. *Behavioral and Brain Sciences* 6, no. 2: 231–245.

Langland-Hassan, P. 2020. *Explaining Imagination.* New York: Oxford University Press.

Lanman, J. A. 2008. In Defence of "Belief": A Cognitive Response to Behaviourism, Eliminativism, and Social Constructivism. *Issues in Anthropology n. s.* 3, no. 3: 49–62.

———. 2020. Belief. *The Immanent Frame: Secularism, Religion, and the Public Sphere* (blog). https://tif.ssrc.org/2020/01/31/belief-lanman/.

Legare, C. H., E. M. Evans, K. S. Rosengren, & P. L. Harris. 2012. The Coexistence of Natural and Supernatural Explanations across Cultures and Development. *Child Development* 83, no. 3: 779–793.

Legare, C. H., & S. A. Gelman. 2008. Bewitchment, Biology, or Both: The Co-Existence of Natural and Supernatural Explanatory Frameworks across Development. *Cognitive Science* 32, no. 4: 607–642.

Legare, C. H., & A. Shtulman. 2018. Explanatory Pluralism across Cultures and Development. In *Metacognitive Diversity: An Interdisciplinary Approach,* edited by J. Proust & M. Fortier, 415–432. Oxford: Oxford University Press.

Leslie, A. M. 1987. Pretense and Representation: The Origins of "Theory of Mind." *Psychological Review* 94, no. 4: 412–426.

———. 1994. ToMM, ToBY, and Agency: Core Architecture and Domain Specificity. In *Mapping the Mind: Domain Specificity in Cognition and Culture,* edited by L. A. Hirschfeld & S. A. Gelman, 119–148. Cambridge: Cambridge University Press.

———. 1995. A Theory of Agency. In *Causal Cognition: A Multidisciplinary Debate,* edited by D. Sperber, D. Premack, and A. J. Premack, 121–149. New York: Oxford University Press.

Levy, N. 2017. Religious Beliefs Are Factual Beliefs: Content Does Not Correlate with Context Sensitivity. *Cognition* 161:109–116.

———. 2018. You Meta Believe It. *European Journal of Philosophy* 26:814–826.

———. 2019. Due Deference to Denialism: Explaining Ordinary People's Rejection of Established Scientific Findings. *Synthese* 196, no. 1: 313–327.

———. 2020. Belie the Belief? Prompts and Default States. *Religion, Brain & Behavior* 10, no. 1: 35–48.

———. 2022. *Bad Beliefs: Why They Happen to Good People.* Oxford: Oxford University Press.

Levy, N., & E. Mandelbaum. 2014. The Powers That Bind: Doxastic Voluntarism and Epistemic Obligation. In *The Ethics of Belief: Individual and Social,* edited by R. Vitz & J. Matheson, 12–33. New York: Oxford University.

Lewis, D. 1973. Counterfactuals. Oxford: Blackwell.

———. 1978. Truth in Fiction. *American Philosophical Quarterly* 15, no. 1: 37–46.

———. 1983. Radical Interpretation. In *Philosophical Papers: Volume I,* 108–118. New York: Oxford University Press.

Lewis, M. 2016. *The Undoing Project: A Friendship That Changed Our Minds.* New York: W. W. Norton.

Lewis, R. 2022. Rolexana. Encyclopedia Britannica. https://www.britannica.com /biography/Roxelana, March 28.

Liao, S. Y., & T. Doggett. 2014. The Imagination Box. *Journal of Philosophy* 111, no. 5: 259–275.

Liberman, Z., A. L. Woodward, & K. D. Kinzler. 2017. The Origins of Social Categorization. *Trends in Cognitive Sciences* 21, no. 7: 556–568.

Lightner, A., & E. H. Hagen. 2022. All Models Are Wrong, and Some Are Religious: Supernatural Explanations as Abstract and Useful Falsehoods about Complex Realities. *Human Nature* 33:425–462.

Lillard, A. S., & D. C. Witherington. 2004. Mothers' Behavior Modifications during Pretense and Their Possible Signal Value for Toddlers. *Developmental Psychology* 40, no. 1: 95–113.

Liquin, E. G., S. E. Metz, & T. Lombrozo. 2020. Science Demands Explanation, Religion Tolerates Mystery. *Cognition* 204, Article 104398.

Lofland, J., & R. Stark. 1965. Becoming a World-Saver: A Theory of Conversion to a Deviant Perspective. *American Sociological Review* 30, no. 6: 862–875.

Lombrozo, T. 2014. Are Factual and Religious Belief the Same? *NPR Cosmos & Culture* (blog). https://www.npr.org/sections/13.7/2014/10/20/357519777 /are-factual-and-religious-belief-the-same, October 20.

Long, T. E., & J. K. Hadden. 1983. Religious Conversion and the Concept of Socialization: Integrating the Brainwashing and Drift Models. *Journal for the Scientific Study of Religion* 22, no. 1: 1–14.

Luhrmann, T. M. 2012. *When God Talks Back: Understanding the American Evangelical Relationship with God.* New York: Vintage.

———. 2018. The Faith Frame: Or, Belief Is Easy, Faith Is Hard. *Contemporary Pragmatism* 15, no. 3: 302–318.

———, ed. 2020a. Mind and Spirit: A Comparative Theory. Special issue. *Journal of the Royal Anthropological Institute* 26, no. S1: 1–166.

————. 2020b. *How God Becomes Real: Kindling the Presence of Invisible Others.*
Princeton, NJ: Princeton University Press.

Luhrmann, T. M., H. Nusbaum, & R. Thisted. 2010. The Absorption Hypothesis:
Learning to Hear God in Evangelical Christianity. *American Anthropologist*
112, no. 1: 66–78.

Luhrmann, T. M., K. Weisman, F. Aulino, J. D. Braninsky, J. C. Dulin, V. A.
Dzkoto, C. H. Legare, M. Lifshitz, E. Ng, N. Ross-Zehnder, & R. E. Smith.
2021. Sensing the Presence of Gods and Spirits across Cultures and Faiths.
Proceedings of the National Academy of Science 118, no. 5, Article e2016649118.

Lyke, M. L. 2003. Church Put to DNA Test. *Seattle Post-Intelligencer,* News,
January 12.

Malhotra, D. K. 2010. When Are Religious People Nicer? Religious Salience and
the "Sunday Effect" on Pro-Social Behavior. *Judgment and Decision Making* 5,
no. 2: 138–143.

Mandelbaum, E. 2014. Thinking Is Believing. *Inquiry: An Interdisciplinary Journal
of Philosophy* 57, no. 1: 55–96.

————. 2019. Troubles with Bayesianism: An Introduction to the Psychological
Immune System. *Mind & Language* 34, no. 2: 141–157.

Mandelbaum, E., & J. Quilty-Dunn. 2015. Believing without Reason, or: Why
Liberals Shouldn't Watch Fox News. *Harvard Review of Philosophy* 22:42–52.

Mark, J. J. 2018. Rollo of Normandy. World History Encyclopedia. https://www
.worldhistory.org/Rollo_of_Normandy/, November 8.

McCauley, R. N. 2011. *Why Religion Is Natural and Science Is Not.* Oxford: Oxford
University Press.

McCauley, R. N., & G. Graham. 2020. *Hearing Voices and Other Matters of the
Mind.* Oxford: Oxford University Press.

McCauley, R. N., & E. T. Lawson. 2002. *Bringing Ritual to Mind: Psychological
Foundations of Cultural Forms.* Cambridge: Cambridge University Press.

McElreath, R., R. Boyd, & P. J. Richerson. 2003. Shared Norms and the Evolution
of Ethnic Markers. *Current Anthropology* 44, no. 1: 122–130.

McGahhey. M, & N. Van Leeuwen. 2018. Interpreting Intuitions. In *Third-Person
Self-Knowledge, Self-Interpretation, and Narrative,* edited by J. Kirsch &
P. Pedrini, 73–98. Switzerland: Springer Nature.

McKay, R. 2004. Hallucinating God? The Cognitive Neuropsychiatry of Religious
Belief and Experience. *Evolution and Cognition* 10, no. 1: 1–10.

McKay, R. T., & R. M. Ross. 2021. Religion and Delusion. *Current Opinion in
Psychology* 40:160–166.

Mele, A. R. 2001. *Self-Deception Unmasked.* Princeton, NJ: Princeton University
Press.

Mercier, H. 2019. *Not Born Yesterday.* Princeton, NJ: Princeton University Press.

Nemeroff, C., & P. Rozin. 2018. Back in Touch with Contagion: Some Essential
Issues. *Journal of the Association for Consumer Research* 3, no. 4: 612–624.

Nisbett, R. E., & L. Ross. 1980. *Human Inference: Strategies and Shortcomings of Social Judgment.* Hoboken, NJ: Prentice Hall.

Norenzayan, A. 2013. *Big Gods: How Religion Transformed Cooperation and Conflict.* Princeton, NJ: Princeton University Press.

Norenzayan, A., & S. Atran. 2004. Cognitive and Emotional Processes in the Transmission of Natural and Nonnatural Beliefs. In *The Psychological Foundations of Culture,* edited by M. Schaller & C. Crandall, 149–169. Hillsdale, NJ: Lawrence Erlbaum.

Norenzayan, A., & A. F. Shariff. 2008. The Origin and Evolution of Religious Prosociality. *Science* 322, no. 5898: 58–62.

Onyibor, M. I. S. 2016. Igbo Cosmology in Chinua Achebe's *Arrow of God*: An Evaluative Analysis. *Open Journal of Philosophy* 6, no. 1: 110–119.

Perry, J. 1980. A Problem about Continued Belief. *Pacific Philosophical Quarterly* 61, no. 4: 317–332.

———. 2001a. *Reference and Reflexivity.* Stanford, CA: CSLI.

———. 2001b. *Knowledge, Possibility, and Consciousness.* Cambridge, MA: MIT Press.

Peterson, M. 2009. *An Introduction to Decision Theory.* Cambridge: Cambridge University Press.

Pierce, J. M. 2017. Why the Catholic Church Bans Gluten-Free Communion Wafers. The Conversation. https://theconversation.com/why-the-catholic -church-bans-gluten-free-communion-wafers-81062, July 21.

Plato. Fourth century BCE/1992. *Platonis Opera.* edited by J. Burnet. Oxford, UK: Oxford Classical Texts.

Porot, N., & E. Mandelbaum. 2021. The Science of Belief: A Progress Report. *WIREs Cognitive Science* 12, no. 2: e1539.

Povinelli, D. 2000. *Folk Physics for Apes: The Chimpanzee's Theory of How the World Works.* Oxford: Oxford University Press.

Qian, S. 94 BCE/2007. *The First Emperor: Selections from the Historical Records,* translated by R. Dawson. Oxford, UK: Oxford World's Classics.

Quattrone, G. A., & A. Tversky. 1984. Causal Versus Diagnostic Contingencies: On Self-Deception and on the Voter's Illusion. *Journal of Personality and Social Psychology* 46, no. 2: 237–248.

Quine, W. V. O. 1960. *Word and Object.* Cambridge, MA: MIT Press.

Ramsey, F. P. 1929/1931. General Propositions and Causality. In *The Foundations of Mathematics and Other Logical Essays,* 237–255. London: Routledge.

Ranney, M. A., & D. Clark. 2016. Climate Change Conceptual Change: Scientific Information Can Transform Attitudes. *Topics in Cognitive Science* 8, no. 1: 49–75.

Raz, J. 1986. *The Morality of Freedom.* New York: Oxford University Press.

Rembold, I. 2017. *Conquest and Christianization.* Cambridge: Cambridge University Press.

Rey, G. 2007. Meta-Atheism: Religious Avowal as Self-Deception. In *Philosophers without Gods,* edited by L. Antony, 243–265. Oxford: Oxford University Press.

Richardson, J. T. 1978. An Oppositional and General Conceptualization of Cult. *Annual Review of the Social Sciences of Religion* 2:29–52.

Roccas, S., L. Sagiv, S. Schwartz, N. Halevy, & R. Eidelson. 2008. Toward a Unifying Model of Identification with Groups: Integrating Theoretical Perspectives. *Personality and Social Psychology Review* 12, no. 3: 280–306.

Rödlach, A. 2006. *Witches, Westerners, and HIV: AIDS & Cultures of Blame in Africa.* Walnut Creek, CA: Left Coast.

Ross, J., & M. Schroeder. 2014. Belief, Credence, and Pragmatic Encroachment. *Philosophy and Phenomenological Research* 88, no. 2: 259–288.

Saribay, S. A., & O. Yilmaz. 2017. Analytic Cognitive Style and Cognitive Ability Differentially Predict Religiosity and Social Conservatism. *Personality and Individual Differences* 114:24–29.

Sartre, J. P. 1943/2003. *Being and Nothingness,* translated by H. E. Barnes. London: Routledge.

Saunders, W. 2014. Disposal of Religious Items. *Arlington Catholic Herald,* May 21.

Sauvayre, R. 2011. Le changement de croyances extremes: du cadre cognitive aux conflits de valeurs. *Revue europeenne des sciences sociales* 49, no. 1: 61–82.

Schellenberg, S. 2013. Belief and Desire in Imagination and Immersion. *Journal of Philosophy* 110, no. 9: 497–517.

Schwitzgebel, E. 2001. In-Between Believing. *Philosophical Quarterly* 51:76–82.

———. 2002. A Phenomenal, Dispositional Account of Belief. *Noûs* 36:249–275.

Shah, N., & J. D. Velleman. 2005. Doxastic Deliberation. *Philosophical Review* 114, no. 4: 497–534.

Shariff, A. F., A. K. Willard, T. Andersen, & A. Norenzayan. 2016. Religious Priming: A Meta-Analysis with a Focus on Prosociality. *Personality and Social Psychology Review* 20, no. 1: 27–48.

Sheikh, H., J. Ginges, & S. Atran. 2013. Sacred Values in the Israeli–Palestinian Conflict: Resistance to Social Influence, Temporal Discounting, and Exit Strategies. *Annals of the New York Academy of Sciences* 1299, no. 1: 11–24.

Sheikh, H., J. Ginges, A. Coman, & S. Atran. 2012. Religion, Group Threat and Sacred Values. *Judgment and Decision Making* 7, no. 2: 110–118.

Shtulman, A. 2013. Epistemic Similarities between Students' Scientific and Supernatural Beliefs. *Journal of Educational Psychology* 105, no. 1: 199–212.

Smith, W. C. 1977/1998. *Believing: An Historical Perspective.* London: Oneworld.

Sosis, R., & C. Alcorta. 2003. Signaling, Solidarity, and the Sacred: The Evolution of Religious Behavior. *Evolutionary Anthropology* 12:264–274.

Spaulding, S. 2018. *How We Understand Others: Philosophy and Social Cognition.* New York: Routledge.

Spelke, E. 1994. Initial Knowledge: Six Suggestions. *Cognition* 50, nos. 1–3: 431–445.

Sperber, D. 1975. *Rethinking Symbolism.* Cambridge: Cambridge University Press.

———. 1982. Apparently Irrational Beliefs. In *Rationality and Relativism,* edited by S. Lukes & M. Hollis, 149–180. Oxford, UK: Blackwell.

———. 1996. *Explaining Culture: A Naturalistic Approach.* Oxford, UK: Blackwell.

———. 1997. Intuitive and Reflective Beliefs. *Mind & Language* 12, no. 1: 67–83.

Stark, R., & R. Finke. 2000. *Acts of Faith: Explaining the Human Side of Religion.* Berkeley: University of California Press.

Stich, S. P. 1981. Dennett on Intentional Systems. *Philosophical Topics* 12, no. 1: 39–62.

Stigall, J. 2018. The Puzzle of Faith. MA thesis, Georgia State University.

Stock, K. 2017. *Only Imagine: Fiction, Interpretation and Imagination.* Oxford: Oxford University Press.

Straus, R. A. 1979. Religious Conversion as a Personal and Collective Accomplishment. *Sociological Analysis* 40, no. 2: 158–165.

Tajfel, H., & J. Turner. 1979. An Integrative Theory of Intergroup Conflict. In *The Social Psychology of Intergroup Relations,* edited by W. G. Austin & S. Worchel, 33–47. Monterey, CA: Brooks/Cole.

Talmont-Kaminski, K. 2013. *Religion as Magical Ideology: How the Supernatural Reflects Rationality.* New York: Routledge.

Tambiah, S. J. 1990. *Magic, Science and Religion and the Scope of Rationality.* Cambridge: Cambridge University Press.

Taves, A. 2016. *Revelatory Events: Three Case Studies of the Emergence of New Spiritual Paths.* Princeton, NJ: Princeton University Press.

Taylor, C. 2007. *A Secular Age.* Cambridge, MA: Harvard University Press.

Taylor, M. 1999. *Imaginary Companions and the Children Who Create Them.* New York: Oxford University Press.

Taylor, M., C. M. Mottweiler, E. R. Naylor, & J. G. Levernier. 2015. Imaginary Worlds in Middle Childhood: A Qualitative Study of Two Pairs of Coordinated Paracosms. *Creativity Research Journal* 27, no. 2: 167–174.

Tetlock, P. 2003. Thinking the Unthinkable: Sacred Values and Taboo Cognitions. *Trends in Cognitive Sciences* 7:320–324.

Tetlock, P. E., O. V. Kristel, S. B. Elson, M. C. Green, & J. S. Lerner. 2000. The Psychology of the Unthinkable: Taboo Trade-Offs, Forbidden Base Rates, and Heretical Counterfactuals. *Journal of Personality and Social Psychology* 78, no. 5: 853–870.

Toren, C. 2007. How Do We Know What Is True? The Case of Mana in Fiji. In *Questions of Anthropology,* edited by R. Astuti, J. P. Parry, & C. Stafford, 307–336. London: Routledge.

Trivers, R. 2011. *Deceit and Self-Deception: Fooling Yourself the Better to Fool Others.* New York: Basic.

Turri, J., D. Rose, & W. Buckwalter. 2018. Choosing and Refusing: Doxastic Voluntarism and Folk Psychology. *Philosophical Studies* 175, no. 10: 2507–2537.

Van Leeuwen, N. 2007. The Spandrels of Self-Deception: Prospects for a Biological Theory of a Mental Phenomenon. *Philosophical Psychology* 20, no. 3: 329–348.

———. 2008. Finite Rational Self-Deceivers. *Philosophical Studies* 139, no. 2: 191–208.

———. 2009a. The Motivational Role of Belief. *Philosophical Papers* 38, no. 2: 219–246.

———. 2009b. Self-Deception Won't Make You Happy. *Social Theory and Practice* 35, no. 1: 107–132.

———. 2010. Why Self-Deception Research Hasn't Made Much Progress. *Philosophy Talk.* (blog) https://www.philosophytalk.org/blog/why-self -deception-research-hasn%E2%80%99t-made-much-progress.

———. 2011. Imagination Is Where the Action Is. *Journal of Philosophy* 108, no. 2: 55–77.

———. 2012. Perry on Self-Knowledge. In *Identity, Language, & Mind: An Introduction to the Philosophy of John Perry,* edited by A. Newen & R. van Riel, 89–107. Stanford, CA: CSLI.

———. 2013. The Meanings of "Imagine" Part I: Constructive Imagination. *Philosophy Compass* 8, no. 3: 220–230.

———. 2014a. Religious Credence Is Not Factual Belief. *Cognition* 133, no. 3: 698–715.

———. 2014b. The Meanings of "Imagine" Part II: Attitude and Action. *Philosophy Compass* 9, no. 11: 791–802.

———. 2016a. Beyond Fakers and Fanatics: A Reply. *Philosophical Psychology* 29, no. 4: 616–621.

———. 2016b. Imagination and Action. In *The Routledge Handbook of Philosophy of Imagination,* 306–319. London: Routledge.

———. 2017a. Two Paradigms for Religious Representation: The Physicist and the Playground (a reply to Levy). *Cognition* 164:206–211.

———. 2017b. Do Religious "Reliefs" Respond to Evidence? *Philosophical Explorations* 20, no. sup1: 52–72.

———. 2018. The Factual Belief Fallacy. *Contemporary Pragmatism* 15, no. 3: 319–343.

———. 2022. Two Concepts of Belief Strength: Epistemic Confidence and Identity Centrality. *Frontiers in Psychology* 13, Article 939949.

Van Leeuwen, N., & T. Lombrozo. 2023. The Puzzle of Belief. *Cognitive Science* 47, no. 2: e13245.

Van Leeuwen, N., & M. van Elk. 2019. Seeking the Supernatural: The Interactive Religious Experience Model. *Religion, Brain & Behavior* 9, no. 3: 221–251.

Van Leeuwen, N., K. Weisman, & T. M. Luhrmann. 2021. To Believe Is Not to Think: A Cross-Cultural Finding. *Open Mind* 5:91–99.

Velleman, J. D. 2000. On the Aim of Belief. In *The Possibility of Practical Reason,* 244–281. New York: Oxford University Press.

Visuri, I. 2020. Sensory Supernatural Experiences in Autism. *Religion, Brain & Behavior* 10, no. 2: 151–165.

Walton, K. L. 1990. *Mimesis as Make-Believe: On the Foundations of the Representational Arts.* Cambridge, MA: Harvard University Press.

Wason, P. C., & P. N. Johnson-Laird. 1972. *Psychology of Reasoning: Structure and Content.* Vol. 86. Cambridge, MA: Harvard University Press.

Watson-Jones, R. E., J. T. Busch, P. L. Harris, & C. H. Legare. 2015. Interdisciplinary and Cross-Cultural Perspectives on Explanatory Coexistence. *Topics in Cognitive Science* 7, no. 4: 611–623.

———. 2017. Does the Body Survive Death? Cultural Variation in Beliefs about Life Everlasting. *Cognitive Science* 41:455–476.

Watson-Jones, R. E., H. Whitehouse, & C. H. Legare. 2016. In-Group Ostracism Increases High-Fidelity Imitation in Early Childhood. *Psychological Science* 27, no. 1: 34–42.

Weatherson, B. 2008. Deontology and Descartes's Demon. *Journal of Philosophy* 105, no. 9: 540–569.

Weisberg, D. S., A. R. Landrum, & J. Hamilton. 2021. Knowledge about the Nature of Science Increases Public Acceptance of Science Regardless of Identity Factors. *Public Understanding of Science* 30, no. 2: 120–138.

Weisberg, D. S., D. M. Sobel, J. Goodstein, & P. Bloom. 2013. Young Children Are Reality-Prone When Thinking about Stories. *Journal of Cognition and Culture* 13, nos. 3–4: 383–407.

Westover, T. 2018. *Educated: A Memoir.* New York: Random House.

Whitehouse, H. 2018. Dying for the Group: Towards a General Theory of Extreme Self-Sacrifice. *Behavioral and Brain Sciences* 41:E192.

Williams, B. 1973. Deciding to Believe. In *Problems of the Self,* 136–151. Cambridge: Cambridge University Press.

Williams, D. 2021. Signalling, Commitment, and Strategic Absurdities. *Mind & Language,* online first.

Wilson, B. R. 1974. *Rationality.* Oxford, UK: Basil Blackwell.

Winch, P. 1964. Understanding Primitive Society. *American Philosophical Quarterly* 1, no. 4: 307–324.

Wittgenstein, L. 1931/1993. *Remarks on Frazer's* Golden Bough. In *Philosophical Occasions* 1912–1951, edited by J. C. Klagge & A. Nordman, 115–155. Indianapolis: Hackett.

Wood, I. 1993. *The Merovingian Kingdoms*: 450–751. New York: Routledge.

Woolley, J. D., & C. A. Cornelius. 2017. Wondering How: Children's and Adults' Explanations for Mundane, Improbable, and Extraordinary Events. *Psychonomic Bulletin & Review* 24, no. 5: 1586–1596.

Xenophanes. Sixth century BCE/1992. *Fragments.* In *Xenophanes of Colophon: Fragments: A Text and Translation with a Commentary,* edited by J. H. Lesher. Toronto: University of Toronto Press.

Zajonc, R. B. 1984. On the Primacy of Affect. *American Psychologist* 39, no. 2, 117–123.

Zimmer, Z., F. Rojo, M. B. Ofstedal, C.-T. Chiu, Y. Saito, & C. Jagger. 2019. Religiosity and Health: A Global Comparative Study. *SSM—Population Health* 7, Article 100322.

Acknowledgments

When I started reading books like this—research monographs—in my first year of college, I looked at the acknowledgment pages with suspicion: surely, I thought, these were exercises in modesty, meant to ingratiate the author to others; all the people listed must have made only a small difference to the final product and nothing of serious substance.

But my former suspicion just shows I didn't really understand how books were written, nor did I realize how much help it takes to get one's ideas worked out well enough to be worth the effort of writing. So, if any college freshmen are reading this (and I hope there are at least a few), know this: the thanks I give here are genuine; nor is this preamble just second-order modesty (philosophers: I see you); and the quality of my final product, such as it is, was improved to some extent by every single person listed here. A few of them greatly improved it.

The project that led to this book first saw light in a 2014 paper that appeared in *Cognition* ("Religious Credence Is Not Factual Belief"). So I first thank my two research assistants who helped me get that paper into shape: Hamza Cherbib and Scott Danielson. I also thank my other research assistants since then who helped with the work that appears here (sometimes with the manuscript, sometimes with connected endeavors, sometimes both): Yanet Berakhi, Keegan Callerame, David Casey, Dawn Chan, Steven Guillemette, Bilal Khan, Katie Lane Kirkland, Casey Landers, Kharli Major, Marlon Rivas Tinoco, Trevor Rukwava, Spencer Smolen, and Hugo Toudic. Hamza and Steven, sadly, are no longer with us. Since they helped make this book a reality, I hope it does its own small part in honoring their memory.

I was also lucky enough to have friends and colleagues who read and commented on the whole manuscript or several key chapters at various stages. Derek Baker, Paul Harris, Neil Levy, Tania Lombrozo, Tanya Luhrmann, Rebecca Tuvel, and Steve Wykstra all stimulated better thought and clearer communication. I recall Tanya telling me with regard to one of the earliest, roughest, and most unnecessarily technically dense sections of a draft of Chapter 1: "It feels like a DO NOT ENTER sign for nonphilosophers." Such comments did wonders in motivating me to improve my writing.

Many others were gracious enough to read, comment on, and/or discuss individual chapters—sometimes regarding matters on which they had notable expertise. For this sort of help, I thank Scott Atran, Olivia Bailey, Dan Dennett, Kenny Easwaran, Larisa Heiphetz, Thad Metz, Daniel Munroe, Eddy Nahmias, Ann Taves, and Evan Westra.

I have also gotten feedback through group discussion. My initial Prologue was greatly improved (shortened!) in light of a helpful round of discussion with the Georgia Tech Philosophy Club. Students in two of my graduate seminars at Georgia State University (GSU) read either the Prologue or individual chapters and helped me see what ideas I needed to communicate more clearly. And I had an especially interesting discussion of Chapter 6 with the students in Juan Piñeros Glasscock's graduate seminar in the spring of 2022. I am also thankful to Amy Kind for including my Chapter 8 in the third/2022 read-ahead online conference on imagination (imaginatively titled: C.O.V.I.D.—Conference [Online/Virtual] for Imagination Domination), and to the participants and my commentators (Josh Myers and Eric Peterson) for a lively discussion.

For other encouragement, discussion, and useful criticism, I thank Wesley Barker, Maarten Boudry, Richard Dub, David Eller, Jonathan Jong, Barry Lam, Eric Mandelbaum, Bob McCauley, Miriam McCormick, Thad Metz, Jim O'Donnell, Eric Schwitzgebel, Azim Shariff, Konrad Talmont-Kaminski, Michiel van Elk, Nicole Vincent, and Kara Weisman. Along the way, I have had many more conversations on the topics of this book than I can count, and these helped strengthen and clarify my thought. I hereby thank the interlocutors I don't name, and do so sincerely.

Many of my views on belief and imagination in general emerged from the process of dissertation writing at Stanford years ago during my PhD work. For help in shaping my central ideas, I thank my advisers Ken Taylor (with whom I would have loved to share this book in his lifetime), John Perry, Krista Lawlor, and Dagfinn Føllesdal. Michael Bratman was also formative in important ways when it comes to how I think about minds and mental states.

Institutional support, for which I'm grateful, came from the following: the European Commission for the Marie Skłodowska-Curie Fellowship I held in the 2015–2016 academic year (call identifier: H2020-MSCA-IF-2014; contract number: 659912) at the Centre for Philosophical Psychology in Antwerp, with the support of its director, Bence Nanay—during this period, I wrote the proposal for this book; the Provost's Office at GSU for the Faculty Fellowship that gave me leave from teaching in the spring of 2018; the Humanities Research Center at GSU for hosting me as a fellow that same spring; and the RISe (Research Intensive Semester) program at GSU for leave from teaching in the fall of 2019. The cross-cultural empirical research reported in Chapter 5 was funded by the John Templeton Foundation through a grant for which Tanya Luhrmann was the PI (Award ID: 55427), and our coauthor, Kara Weisman, was also funded by the National Science Foundation for part of the time we were collaborating on that research (Award ID: DGE-114747). All these institutions deserve some measure of credit.

I thank my editor Andrew Kinney for his help in shaping the proposal that was the guiding template for this book and for his careful and insightful comments on everything I wrote, right down to these very acknowledgments. Two anonymous referees for Harvard University Press also did a splendid job of finding ways to improve the manuscript; I am grateful to them as well. In their honor, I will occasionally buy a drink for a complete stranger.

My father, Raymond Van Leeuwen, is a biblical scholar, and my mother, Mary Stewart Van Leeuwen, is a philosophically oriented academic psychologist. They have read some of my work over the years and cheered it on, but they haven't seen the writing in this book until now. Still, I must thank them for making my brain the sort of field where the right seeds could grow into projects like this one. Given that intellectual provenance, I can only say to my readers: Well, what did you expect?

* * *

Finally, I hope that many who see the title of this book will recognize its homage to Kendall Walton's stunning theoretical work *Mimesis as Make-Believe: On the Foundations of the Representational Arts*. I hope even more that I have worked well enough with the ideas and (even more importantly) theoretical approach in that book to deserve to be paying it such homage. The ghosts of the play we do as children linger in the cultural forms that shape our social worlds as adults. To understand religion, we must make the ghosts speak.

Index